DATE DUE			
Nov 3 '76			
Apr 18 77			
Dec 13 77			
Oct 25 '82			

Chromosomal
Proteins and Their
Role in the Regulation of
Gene Expression

Colloquium Organizing Committee

Gary Stein, *Chairman*
Lewis Kleinsmith, *Co-chairman*
Peter Cerutti
Rusty Mans
Philip Laipis
Eugene Sander

Proceedings of the
Florida Colloquium on Molecular Biology
March 13-14, 1975
Sponsored by the Department of Biochemistry
University of Florida
Gainesville, Florida

Chromosomal Proteins and Their Role in the Regulation of Gene Expression

Edited by

GARY S. STEIN
Department of Biochemistry
The University of Florida
Gainesville, Florida

LEWIS J. KLEINSMITH
Department of Zoology
The University of Michigan
Ann Arbor, Michigan

 Academic Press, Inc.

New York   San Francisco   London   1975
A Subsidiary of Harcourt Brace Jovanovich Publishers

574.88

F66c

97588

July 1976

ACADEMIC PRESS, INC.
111 Fifth Avenue, New York, New York 10003

United Kingdom Edition published by
ACADEMIC PRESS, INC. (LONDON) LTD.
24/28 Oval Road, London NW1

Library of Congress Cataloging in Publication Data

Florida Colloquim on Molecular Biology, University of
        Florida, Gainesville, 1975.
        Chromosomal proteins and their role in the regulation
of gene expression.

        "Proceedings of the Florida Colloquim on Molecular
Biology, March 13-14, 1975."
        Includes bibliographies and index.
        1.   Molecular biology–Congresses.   I.   Stein, Gary S.
II.   Kleinsmith, Lewis J.   III.   Florida. University,
Gainesville.   Dept. of Biochemistry.   [DNLM:
1.   Genetics, Biochemical–Congresses.   2.   Chromosomes–
Physiology–Congresses.   3.   Chromatin–Physiology–
Congresses.   4.   Proteins–Physiology–Congresses.
QH450 F636c 1975]
QH506.F55 1975   574.8'8   75-22181
ISBN 0–12–664750–X

Contents

CONTENTS

Contributors

Vincent G. Allfrey, The Rockefeller University, New York, New York 10021

Raghuveera Ballal, Nuclear Protein and Tumor By-Products Laboratories, Department of Pharmacology, Baylor College of Medicine, Houston, Texas 77025

C. Stuart Baxter, Department of Pathology, University of Florida, College of Medicine, Gainesville, Florida 32610

James Bonner, Division of Biology, California Institute of Technology, Pasadena, California 91125

Harris Busch, Nuclear Protein and Tumor By-Products Laboratories, Department of Pharmacology, Baylor College of Medicine, Houston, Texas 77025

Rose K. Busch, Nuclear Protein and Tumor By-Products Laboratories, Department of Pharmacology, Baylor College of Medicine, Houston, Texas 77025

Paul Byvoet, Department of Pathology, University of Florida, College of Medicine, Gainesville, Florida 32610

Roger Chalkley, Department of Biochemistry, The University of Iowa, Iowa City, Iowa 52242

Edward Ezrailson, Nuclear Protein and Tumor By-Products Laboratories, Department of Pharmacology, Baylor College of Medicine, Houston, Texas 77025

Carl M. Feldherr (Session Chairman), Department of Pathology, University of Florida, Gainesville, Florida 32610

Gary Felsenfeld, Laboratory of Molecular Biology, National Institute of Arthritis, Metabolism, and Digestive Diseases, Bethesda, Maryland 20014

R. Stewart Gilmour, The Beatson Institute for Cancer Research, Glasgow, Scotland G3 6UD

Ira L. Goldknopf, Nuclear Protein and Tumor By-Products Laboratories,

Department of Pharmacology, Baylor College of Medicine, Houston, Texas 77025

Joel Gottesfeld, Division of Biology, California Institute of Technology, Pasadena, California 91125

Daryl K. Granner, Department of Biochemistry, The University of Iowa, Iowa City, Iowa 52242

Sidney R. Grimes, Department of Biochemistry, University of Texas System Cancer Center, M. D. Anderson Hospital and Tumor Institute, Houston, Texas 77025

Lubomir S. Hnilica, Department of Biochemistry, School of Medicine, Vanderbilt University, Nashville, Tennessee 37232

Philip Hohmann, Biomedical Research Group, H-9, Los Alamos Scientific Laboratory, Los Alamos, New Mexico 87544

Gwen Hord, Department of Biochemistry, University of Texas System Cancer Center, M. D. Anderson Hospital and Tumor Institute, Houston, Texas 77025

Akira Inoue, The Rockefeller University, New York, New York 10021

Vaughn Jackson, Department of Biochemistry, The University of Iowa, Iowa City, Iowa 52242

Edward M. Johnson, The Rockefeller University, New York, New York 10021

Diane Kent, Division of Chemistry and Chemical Engineering, California Institute of Technology, Pasadena, California 91125

Lewis J. Kleinsmith (Session Chairman), Department of Zoology, University of Michigan, Ann Arbor, Michigan 48104

Nina C. Kostraba, Division of Cell and Molecular Biology, State University of New York at Buffalo, Buffalo, New York 14214

Thomas A. Langan, Department of Pharmacology, University of Colorado Medical School, Denver, Colorado 80220

Rusty J. Mans (Session Chairman), Department of Biochemistry, University of Florida, Gainesville, Florida 32610

Marvin L. Meistrich, Department of Experimental Radiotherapy, University of Texas System Cancer Center, M. D. Anderson Hospital and Tumor Institute, Houston, Texas 77025

Mark O. J. Olson, Nuclear Protein and Tumor By-Products Laboratories,

Department of Pharmacology, Baylor College of Medicine, Houston, Texas 77025

William Park, Department of Biochemistry, University of Florida, Gainesville, Florida 32610

Gordhan L. Patel, Department of Zoology, University of Georgia, Athens, Georgia 30602

John Paul, The Beatson Institute for Cancer Research, Glasgow, Scotland G3 6UD

George M. Pikler, Department of Molecular Medicine, Mayo Clinic, Rochester, Minnesota 55901

Robert D. Platz, Department of Biochemistry, University of Texas System Cancer Center, M. D. Anderson Hospital and Tumor Institute, Houston, Texas 77025

Archie W. Prestayko, Nuclear Protein and Tumor By-Products Laboratories, Department of Pharmacology, Baylor College of Medicine, Houston, Texas 77025

Michael Ross, Division of Chemistry and Chemical Engineering, California Institute of Technology, Pasadena, California 91125

Barbara Sollner–Webb, Laboratory of Molecular Biology, National Institute of Arthritis, Metabolism, and Digestive Diseases, Bethesda, Maryland 20014

Thomas C. Spelsberg, Department of Molecular Medicine, Mayo Clinic, Rochester, Minnesota 55901

Gary S. Stein (Session Chairman), Department of Biochemistry, University of Florida, Gainesville, Florida 32610

Janet L. Stein, Department of Biochemistry, University of Florida, Gainesville, Florida 32610

Charles W. Taylor, Nuclear Protein and Tumor By-Products Laboratories, Department of Pharmacology, Baylor College of Medicine, Houston, Texas 77025

Terry L. Thomas, Department of Zoology, University of Georgia, Athens, Georgia 30602

Cary Thrall, Department of Biochemistry, University of Florida, Gainesville, Florida 32610

Tung Yue Wang, Division of Cell and Molecular Biology, State University of New York at Buffalo, Buffalo, New York 14214

Robert Webster, Department of Molecular Medicine, Mayo Clinic, Rochester, Minnesota 55901

Lynn C. Yeoman, Nuclear Protein and Tumor By-Products Laboratories, Department of Pharmacology, Baylor College of Medicine, Houston, Texas 77025

Preface

It was not long ago that considerations of the eukaryotic genome focused almost entirely on DNA, whereas chromosomal proteins were only briefly mentioned or at times even ignored. In recent years the situation has changed dramatically. Intensive efforts have now been directed toward examining the proteins associated with the genome, with initial attention being focused on the histones followed by a growing concern with the nonhistone chromosomal proteins. It is becoming increasingly apparent that these classes of protein may dictate important structural as well as transcriptional properties of the genetic material.

The existence of histones in nuclei of eukaryotic cells was documented prior to the turn of the century, and an involvement of these basic chromosomal proteins in the regulation of gene expression was postulated as early as three decades ago. However, it has only been during the past 15 years that intensive experimental efforts employing metabolic and structural probes have yielded convincing evidence that histones play important roles in both restricting the transcriptional capacity of DNA and in the packaging of the genetic information. Despite the level to which our understanding of the structural and functional properties of the histones has progressed, many important problems remain to be resolved. The precise manner in which histones interact with other genome components requires clarification. The interrelationship of histone synthesis and DNA replication must be further elucidated. The distribution of histones along active and inactive regions of the DNA molecule is not known. Answers to these questions and many others are undoubtedly forthcoming.

Although a structural and regulatory role for the histones has been clearly documented, it has become increasingly apparent that these proteins lack the specificity for recognition of individual genetic loci. As a result, attention has recently turned to the nonhistone protein fraction as a potential source of specific regulators of gene readout, i.e., molecules capable of interacting with defined DNA sequence in such a manner as to render them more or less transcribable. These proteins originally were described as early as the 1940s, but their insolubility and tendency to form aggregates has presented a major obstacle to their characterization. This problem was partially circumvented in the late 1960s with the development of SDS polyacrylamide gel electrophoretic fractionation methods, which afforded an opportunity for separating the nonhistone chromosomal proteins in a soluble (although denatured) state. As a result, it has been possible to demonstrate in numerous biological systems that variations in the composition and metabolism of nonhistone chromosomal proteins are correlated with modifications in gene expression. Some of the nonhistone proteins have also been shown to be capable of specific binding to DNA. Both of these types of observations support the notion that these proteins are in some way involved in the regulation of transcription. More direct evidence has come from chromatin reconstitution experiments, which have demonstrated that components of the nonhistone chromosomal proteins mediate control of transcription of tissue-specific and cell cycle stage-specific genes. Although it is undeniable that we have made definitive inroads toward elucidating the involvement of nonhistone chromosomal proteins in the regulation of transcription, we have merely uncovered the first set of clues to an extremely complex problem. Perhaps the most obvious problem is identifying which among the highly heterogeneous nonhistone chromosomal proteins are the specific regulatory proteins. Another series of problems concerns the distribution of the nonhistone chromosomal proteins in chromatin and the mechanism by which they render genetic sequences transcribable.

This book resulted from the 1975 *Florida Colloquium on Molecular Biology*, the topic of which was "Chromosomal Proteins and Their Role in the Regulation of Gene Expression." The colloquium, and hence the book, considered many current approaches for studying the structural and functional properties of chromosomal proteins, particularly as they relate to the control of transcription. Predicated upon the rapid progress in this area during the past several years and the recent development of high resolution probes for further work, it is reasonable to anticipate that problems previously unresolvable can now be meaningfully pursued.

The editors are indebted to Janet Stein for her editorial assistance and to Bonnie Cooper for typing the manuscripts.

REGULATION OF HISTONE GENE TRANSCRIPTION DURING THE CELL CYCLE BY NONHISTONE CHROMOSOMAL PROTEINS

Gary Stein, Janet Stein, Cary Thrall and William Park
Department of Biochemistry
University of Florida
Gainesville, Florida

Abstract

We have examined the regulation of histone gene expression during the cell cycle and the role of nonhistone chromosomal proteins in such regulation. Adenylic acid residues were added to the 3'-OH ends of histone messenger RNAs, and using RNA-dependent DNA polymerase, a complementary DNA (cDNA) was synthesized. This high resolution probe has been utilized to demonstrate that: a. Histone messenger RNA sequences are present in the polyribosomal RNA of S phase but not of G_1 cells; b. Transcription in vitro of histone messenger RNA sequences is restricted to the S phase of the cell cycle; and c. Nonhistone chromosomal proteins are responsible for regulating the transcription of those regions of the genome which contain the information for the synthesis of histones.

The cell cycle of continuously-dividing cells provides an effective model system for studying the regulation of gene expression. Many of the complex and interdependent biochemical events occurring throughout the cell cycle require the differential elaboration of information contained within the genome (1,2). The onset of DNA replication and mitosis are both essential events for which RNA synthesis is an absolute requirement, indicating that their regulation resides, at least in part, at the transcriptional level. An important property of the cell cycle is that it offers the opportunity to study the mechanism by which the transient expression of genes is regulated. This is in contrast to the control of genes which are expressed on a more permanent basis, as in differentiated systems such as erythroid cells which are committed to the synthesis of hemoglobin.

Several lines of evidence suggest that nonhistone chromosomal proteins play a key role in the regulation of gene expression in general (1,3-7) and, specifically, in the control of

1

transcription during the cell cycle (1,8,9). Variations in the
composition and metabolism of nonhistone chromosomal proteins
have been reported throughout the cell cycle of continuously-
dividing cells (10-16). Such cell cycle stage-specific variations in nonhistone chromosomal proteins are consistent with
the hypothesis that amongst these proteins are regulatory
macromolecules. However, these data are of a correlative nature and, hence, the evidence is at best circumstantial. A
more direct assessment of the involvement of nonhistone chromosomal proteins in the regulation of transcription during the
cell cycle is provided by a series of chromatin reconstitution
studies. In these studies, it was shown that the nonhistone
chromosomal proteins are responsible for the differences in
template activity of S phase and mitotic chromatin (8).

The present article is concerned with the role of nonhistone chromosomal proteins in regulating the transcription of
histone genes. These genetic sequences are transiently expressed during a restricted period of the cell cycle and are
of particular significance since their expression is function-
ally related to DNA replication. Furthermore, the gene products, histones, are themselves of key importance in the regulation of transcription and in the maintenance of genome
structure.

In Vitro Synthesis of DNA Complementary to Histone Messenger RNA

A high resolution probe for identification and quantita-
tion of histone mRNA sequences, both associated with polyribosomes and as components of RNA transcripts, is essential for
critically studying the regulation of histone gene expression.
We therefore synthesized a single-stranded DNA complementary to
histone messenger RNAs. The approach pursued for the in vitro
synthesis of a complementary histone DNA is schematically illustrated in Figure 1.

7-12S RNAs were isolated from the polyribosomes of S phase
HeLa S_3 cells by successive sucrose gradient fractionations.
To eliminate poly A-containing RNAs which might sediment in the
7-12S region of the sucrose gradients, the 7-12S RNAs were further fractionated by passage through nitrocellulose (Millipore)
filters (17) or by affinity chromatography on oligo dT-cellu-
lose under conditions where poly A-containing material is retained (18). Greater than 95% of the RNA is recovered after
either fractionation, and the size distribution of the 7-12S
RNA on SDS polyacrylamide gels is shown in Figure 2. An apparent heterogeneity is evident with 4 defined peaks. Based on

2

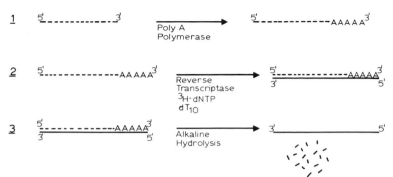

Fig. 1 (above). Caption on following page.

Fig. 2 (below). Caption on following page.

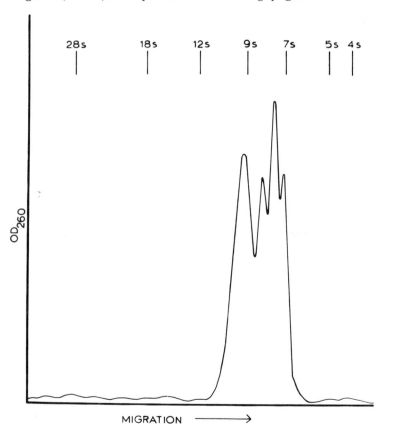

the molecular weights of the individual histone proteins, it is reasonable to anticipate that each of the peaks is enriched in the mRNAs for specific histones.

To confirm that the 7-12S phase polysomal RNAs comprise the mRNAs for histones, they were translated in cell-free protein-synthesizing systems derived from rabbit reticulocyte (19) and wheat germ (20) lysates. In both systems these RNAs stimulated the synthesis of F_{2a1}, F_{2a2}, F_{2b}, F_3 and F_1 histone polypeptides, as evidenced by their extractability with dilute mineral acid (0.4 N H_2SO_4) and by their co-electrophoresis with standard HeLa S_3 cell histone markers in high resolution acetic acid-urea polyacrylamide gels. The electrophoretic fractionation of histones synthesized in a wheat germ lysate using 7-12S RNA from S phase HeLa S_3 cells as template is shown in Figure 3.

Poly A sequences are required at the 3'-OH termini of messenger RNA molecules to complex with oligo dT, which serves as an effective primer for RNA-dependent DNA polymerase. Histone mRNAs as isolated from polyribosomes of S phase HeLa S_3 cells lack extensive sequences of poly A at their 3'-OH termini. Adenylic acid residues were therefore enzymatically added to the 3'-OH ends of histone mRNAs in a covalent 3',5'-phosphodiester linkage with an ATP-polynucleotidylexotransferase isolated from maize seedlings. Polyadenylation of histone mRNA was carried out in collaboration with Dr. Rusty J. Mans (Department of Biochemistry, University of Florida, Gainesville, Florida). The isolation and purification of the enzyme has been reported (21). By manipulating the reaction conditions, principally the ratio of enzyme:mRNA and the time of incubation, the length of the poly A tails on the histone mRNA can be varied. We have found that the optimal extent of polyadenyla-

---

Fig. 1. Schematic illustration of protocol for histone cDNA synthesis. 1. Adenylic acid residues are enzymatically added to the 3'-OH termini of histone mRNAs utilizing an ATP-polynucleotidyl exotransferase. 2. The polyadenylated histone mRNAs are transcribed with RNA-dependent DNA polymerase in the presence of ^3H-dGTP and ^3H-dCTP using dT_{10} as a primer. (^3H)-DNA-polyadenylated histone mRNA duplexes are formed. 3. Alkaline hydrolysis of the polyadenylated histone mRNA component of the DNA-RNA duplex results in a ^3H-single-stranded DNA complementary to histone mRNAs. Details concerning the synthesis of the histone cDNA probe have been reported (33).

Fig. 2. Polyacrylamide gel electrophoretic profile of histone mRNAs isolated from the polyribosomes of S phase HeLa S_3 cells.

4

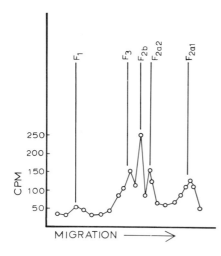

Fig. 3. Electrophoretic fractionation of ^{3}H-leu-labeled
histone polypeptides synthesized from the 7-12S polyribosomal RNAs of S phase HeLa S_3
cells in a cell-free protein-
synthesizing system derived
from wheat germ. The radioactivity incorporated in the
absence of 7-12S HeLa RNA has
been subtracted. Electrophoresis was carried out in acetic
acid-urea polyacrylamide gels
in the presence of purified
histone standards as indicated.

tion for transcription of a complementary DNA is approximately
35 AMP residues per molecule of histone message. The average
chain length of poly A added to histone mRNAs was calculated
from the AMP:adenosine ratio after the polyadenylated mRNA
preparation was subjected to alkaline hydrolysis.

A single-stranded ^{3}H-DNA complementary to the polyadenylated histone messenger RNAs was synthesized using RNA-dependent
DNA polymerase (reverse transcriptase) isolated from Rous sar-
coma virus with dT_{10} as a primer. The reaction was carried out
in the presence of actinomycin D to insure that the DNA copy
was single-stranded (22). Since the cDNA is greater than 96%
sensitive to S_1 nuclease (a single-strand specific nuclease
(23) which utilizes DNA as well as RNA as a substrate), it can
be concluded that little, if any, double-strandedness is present in the histone cDNA probe. Initially, the minimum length
of the bulk of the cDNA was determined to be at least 200 nucleotides. This minimum length for cDNA was based on exclusion
from Sephadex G-100 when chromatographed in the presence of 8M
urea on a column calibrated for deoxynucleotide polymers. The
actual size of the cDNA as measured by sedimentation in alkaline sucrose gradients is approximately 400 nucleotides, based
on a mean sedimentation coefficient of 6.1S.

The kinetics of hybridization of the cDNA probe to histone
mRNA isolated from polysomes of S phase HeLa S_3 cells were
measured to assess the purity of the template RNA and the cDNA
(Figure 4). The hybridization reaction was carried out in the
presence of 50% formamide and 0.5 M NaCl. We determined the

5

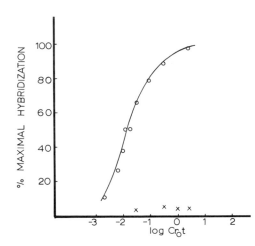

Fig. 4. Kinetics of annealing of histone cDNA to histone messenger RNA isolated from the polysomes of S phase HeLa S_3 cells. 0.37 ng of ^{3}H-cDNA (27,000 dpm/ng) was annealed at 52° in a volume of 15 μl containing 50% formamide, 0.5 M NaCl, 25 mM HEPES (pH 7.0) and 1 mM EDTA with either 0.03 or 0.19 μg of histone mRNA in the presence of 3.75 μg of E. coli RNA as carrier (ooo) or with 3.75 μg of E. coli RNA under identical conditions (xxx). No background values have been subtracted. $C_{r_o}t$ = mole ribonucleotides x sec/liter.

Fig. 5. Rate of histone messenger RNA-cDNA hybridization as a function of temperature. 0.19 μg of histone mRNA was annealed with 0.37 ng of ^{3}H-cDNA for 30 minutes ($C_{r_o}t$ = 6.25 x 10^{-2}). Each reaction mixture also contained 3.75 μg of E. coli RNA.

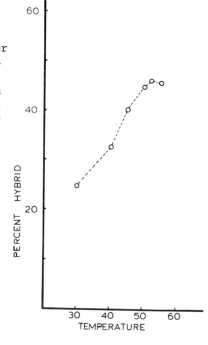

optimal temperature for the annealing reaction by measuring the extent of hybrid formation between histone mRNA and cDNA at a C_rt of 6.25×10^{-2}. The temperature-dependent rate curve shown in Figure 5 indicates that under these conditions the rate of the hybridization reaction is maximal at 52°. It should be noted that these conditions are extremely stringent since the temperature is only 13°C below the T_m of histone mRNA-cDNA hybrids (65°C). The $C_{r_0}t_{1/2}$ of the hybridization reaction at 52° is 1.7×10^{-2}. Since the molar concentration of the nucleotide sequences in solution determines the rate of hybridization, comparison of the $C_{r_0}t_{1/2}$ of the histone mRNA to that of a kinetic standard such as globin mRNA (complexity of 1200 bases (24) and $C_{r_0}t_{1/2}$ under similar conditions of 3.8 $\times 10^{-3}$(25)) yields a calculated sequence complexity of approximately 5400 bases which is 2 times greater than expected for the total complexity of the 5 histone messages. However this is within the range of variation found for the rate of RNA-DNA hybridization (26-29). When cDNA and histone mRNA are hybridized in the absence of formamide at 75°C, the $C_{r_0}t_{1/2}$ is 5 x 10^{-3}. From the $C_{r_0}t_{1/2}$ of globin mRNA under these conditions ($C_{r_0}t_{1/2}$ = 2.0×10^{-3}) (30,31) the sequence complexity of histone mRNA can be estimated to be 3000 bases. When the probe is annealed with E. coli RNA under either of the above conditions, no significant level of hybrid formation above background is detected (Figure 4). Thermal denaturation curves of the histone mRNA-cDNA hybrids exhibit a single transition with a T_m of 65°C in 50% formamide-0.5 M NaCl-25 mM HEPES (pH 7.0)-1 mM EDTA and 95°C in 0.5 M NaCl-25 mM HEPES (pH 7.0)-1 mM EDTA. These T_m values are consistent with the reported base composition of histone mRNA of 54% G-C (32). The preparation and properties of the histone cDNA probe are described in a recent publication from this laboratory (33).

Hybridization Analysis of Histone Messenger RNA Association with Polyribosomes During the Cell Cycle

A functional relationship between histone synthesis and DNA replication is suggested by the fact that in many biological systems the synthesis of these proteins and their deposition on the DNA is restricted to the S phase of the cell cycle (10,34,35). Further support for the coupling of histone and DNA synthesis comes from the observation that inhibition of DNA replication results in a rapid shutdown of histone synthesis (10,35-37). It has previously been shown, utilizing cell-free protein-synthesizing systems derived from reticulocytes and Ehrlich ascites cells, that the RNA isolated from polyribosomes of S phase HeLa cells supports the synthesis of histones while the RNA from polyribosomes of G_1 phase cells or of S phase cells

7

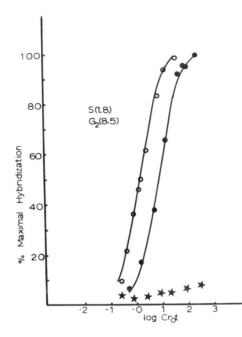

Fig. 6. ^{3}H–cDNA (27,000 dpm/ng) and unlabelled RNA were hybridized at 52°C in sealed glass capillary tubes containing in a volume of 15 µl: 50% formamide, 0.5 M NaCl, 25 mM HEPES (pH 7.0), 1 mM EDTA, 0.37 ng cDNA and 3.75 or 7.5 µg of polysomal RNA from G_1 (✶ ✶ ✶), S (oooo) or G_2 (●●●●) HeLa S_3 cells. Samples were removed at various times and incubated for 20 min in 2.0 ml of 30 mM sodium acetate, 0.3 M NaCl, 1 mM $ZnSO_4$, 5% glycerol (pH 4.6) containing S_1 nuclease at a concentration sufficient to degrade at least 96% of the single-stranded nucleic acids present. The amount of radioactive DNA resistant to digestion was determined by trichloroacetic acid precipitation. S and G_2 phase cells were obtained by synchronization with 2 cycles of 2 mM thymidine block. S phase cells were harvested 3 hours after release from the second thymidine block at which time 98% of the cells were in S phase. G_2 cells were harvested 7.5 hours after release from thymidine. G_1 cells were obtained 3 hours after selective detachment of mitotic cells from semi-confluent monolayers; 97% of the cells were in the G_1 phase of the cell cycle and S phase cells were not detected. Polyribosomal RNA was isolated as reported (49).

treated with inhibitors of DNA synthesis does not (19,38,39). These findings indicate that translatable histone mRNAs are associated with polyribosomes exclusively during the S phase of the cell cycle. However, the possibility still exists that histone mRNAs are components of the polyribosomes during other periods of the cell cycle, but have in some way been rendered nontranslatable. Such a possibility would have important implications for the mechanism operative in the regulation of histone gene expression. Therefore, to establish that histone mRNA sequences are associated with polyribosomes only during S phase, we examined G_1, S and G_2 polyribosomal RNAs for their ability to hybridize with histone cDNA.

The extent of hybrid formation between histone cDNA

and total polysomal RNA of G_1, S and G_2 cells is compared in Figure 6. The hybridization observed between S phase polyribosomal RNA and the cDNA indicates the presence of histone-specific sequences associated with the polyribosomes of S phase cells. In contrast the absence of G_1 polyribosomal RNA hybridization demonstrates that histone mRNA sequences are not components of G_1 polyribosomes. A comparison of the kinetics of the hybridization reaction between S phase polyribosomal RNA and cDNA ($C_{ro}t_{1/2}$ = 1.8) with the hybridization kinetics of the histone mRNA-cDNA reaction ($C_{ro}t_{1/2}$ = 1.7 x 10^{-2}) indicates that histone mRNA sequences account for 0.9% of the S phase total polysomal RNA. This figure is consistent with the in vivo situation where approximately 10% of the protein synthesis during S phase is histone synthesis (10). Additionally the complete absence of hybrid formation between G_1 polysomal RNA and histone cDNA establishes the absence of ribosomal RNA (5S, 18S and 28S) and tRNA sequences in the histone mRNA preparation as well as in the cDNA probe.

Determination of the presence or absence of histone mRNA sequences on G_2 polysomes is complex. The kinetics of the hybridization reaction between G_2 polyribosomal RNA and histone cDNA ($C_{ro}t_{1/2}$ = 8.5) suggests that the amount of histone mRNA sequences present on the polyribosomes of G_2 cells is 21% of that present on S phase polyribosomes. However, the data in Figure 7 clearly indicate that 20% of the G_2 cell population consists of cells which are undergoing DNA replication (S phase cells). It is therefore reasonable to conclude that the histone mRNA sequences present in the G_2 polyribosomal RNA are due to the presence of S phase cells in the G_2 cell population. This implies that histone mRNA sequences are not associated with polyribosomes during the G_2 phase of the cell cycle. Unfortunately, to date, no effective methodology is available for obtaining a pure population of G_2 phase HeLa S_3 cells to establish this point definitively.

These results indeed demonstrate that in HeLa cells histone mRNA sequences are associated with polyribosomes only during the S phase of the cell cycle (43). It therefore follows that regulation of histone gene expression in this system does not reside at the translational level and transcriptional control is strongly implied. However, this type of regulation of histone gene expression may not be universal. For example, there is evidence that during early stages of embryonic development, control of histone synthesis may be mediated, at least in part, post-transcriptionally (40,41).

9

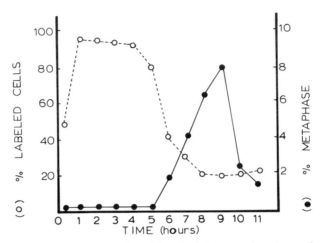

Fig. 7. Percentage of cells in DNA synthesis and mitotic index at various times following release of HeLa S_3 cells from 2 cycles of 2 mM thymidine block. Cells were labelled with 5 μCi/ml of thymidine ^3H for 15 minutes and the percentage of cells in DNA synthesis was determined autoradiographically. The mitotic index was determined from the autoradiographic preparations.

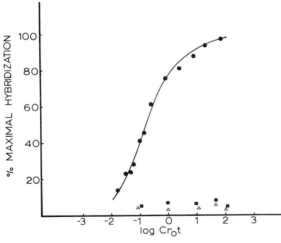

Fig. 8. Kinetics of annealing of histone cDNA to in vitro transcripts of chromatin from G_1 and S phase HeLa S_3 cells. 0.37 ng of ^3H-cDNA (27,000 dpm/ng) was annealed at 52° to either 0.15 or 1.5 μg of RNA transcripts from G_1 (△ △ △) or S phase (ooo) chromatin. 0.37 ng of cDNA was also annealed to 1.5 μg of E. coli RNA isolated in the presence of S phase chromatin (■ ■). E. coli RNA was included in each reaction mixture so that the final amount of RNA was 3.75 μg.

Cell Cycle Stage-Specific Transcription of Histone Genes

To ascertain directly that the genes which contain the information for histone synthesis are transcribed during a restricted period of the cell cycle, the following approach was pursued. Chromatin from G_1 and S phase cells was transcribed in a cell-free system, the RNAs were isolated and their ability to form S_1 nuclease-resistant, TCA-precipitable hybrids with histone cDNA was determined. The kinetics of hybridization of histone cDNA and RNA transcripts from G_1 as well as S phase chromatin are shown in Figure 8. While transcripts from S phase chromatin hybridize with histone cDNA at a $C_{r_0}t_{1/2}$ of 2 x 10^{-1} compared with a $C_{r_0}t_{1/2}$ of 1.7 x 10^{-2} for the histone mRNA-cDNA hybridization reaction, there is no evidence of hybrid formation between histone cDNA and G_1 transcripts, even at a $C_{r_0}t$ of 100. The maximal level of hybrid formation between histone cDNA and S phase transcripts was the same as that observed between histone cDNA and histone mRNA. Fidelity of the hybrids formed between histone cDNA and transctipts from S phase chromatin is suggested by the fact that the T_m of these hybrids is identical to the T_m of histone mRNA-cDNA hybrids ($65°$). It should be noted that the T_m obtained under these conditions is consistent with an RNA-DNA hybrid having a G-C content of 54%, which is the nucleotide composition of histone mRNA reported by Adesnik and Darnell (32).

RNAs synthesized in intact cells may remain associated with chromatin during isolation and in part account for hybrid formation between in vitro RNA transcripts and complementary DNAs for specific genes. It is possible that the extent to which this phenomenon occurs varies significantly with the tissue or cell and the method of chromatin preparation. To determine if such endogenous RNA sequences account for histone specific sequences which are detected in transcripts from S phase chromatin, the following control was executed. S phase chromatin was placed in the in vitro transcription mixture without RNA polymerase and an amount of E. coli RNA equivalent to the amount of RNA transcribed from S phase chromatin was added. RNA was immediately extracted by the same procedure utilized for the isolation of in vitro RNA transcripts. When this control RNA was annealed with histone cDNA, no significant level of hybridization was observed (Figure 8). These results establish that endogenous histone-specific sequences associated with S phase chromatin are not contributing significantly to the hybridization observed with S phase in vitro transcripts. It is therefore reasonable to conclude that the histone sequences present in S phase transcripts can be totally accounted for by in vitro synthesis.

11

The results from these in vitro transcription studies
clearly indicate that histone sequences are available for transcription during S phase and not during G_1 (42). Such findings
are consistent with the restriction of histone synthesis to the
S phase of the cell cycle (10,34,35) and the association of
histone messenger RNAs with polysomes only during S phase (36-
39). Taken together, this evidence suggests that in continu-
ously dividing HeLa S_3 cells the expression of histone genes
is regulated at the transcriptional level and that the readout
of these genetic sequences occurs only during the period of
DNA replication.

Regulation of Histone Gene Transcription by Nonhistone Chromosomal Proteins

Although evidence has been presented which strongly suggests that amongst the nonhistone chromosomal proteins are
macromolecules which are responsible for the regulation of
transcription during the cell cycle (1,2,8-16), the evidence
is primarily of a correlative nature. To examine directly the
involvement of nonhistone chromosomal proteins in the control
of cell cycle stage-specific gene readout, we pursued the following approach. Chromatin isolated from G_1 and S phase cells
was dissociated, fractionated, and reconstituted as outlined
in Figure 9. In vitro RNA transcripts from chromatin reconstituted with G_1 nonhistone chromosomal proteins and from chromatin reconstituted with S phase nonhistone chromosomal proteins
were annealed with histone ^{3}H-cDNA. Figure 10 indicates that
RNA transcripts from chromatin reconstituted with S phase nonhistone chromosomal proteins hybridize with histone cDNA while
those from chromatin reconstituted with G_1 nonhistone chromosomal proteins do not exhibit a significant degree of hybrid
formation. It should be emphasized that the kinetics and extent of hybridization with the cDNA are the same for transcripts of native S phase chromatin and transcripts of chromatin reconstituted with S phase nonhistone chromosomal proteins
(Figure 10). Furthermore, the amounts of RNA transcribed and
the recoveries during isolation of these transcripts from native and reconstituted chromatin preparations are essentially
identical. These results clearly imply a functional role for
nonhistone chromosomal proteins in regulating the availability
of histone sequences for transcription during the cell cycle
(49). Such a regulatory role for nonhistone chromosomal proteins is in agreement with results from several laboratories
which have indicated that these proteins are responsible for
the tissue-specific transcription of globin genes (30,44, and
L. S. Hnilica, personal communication). However, the present
results represent the first demonstration that nonhistone chro-

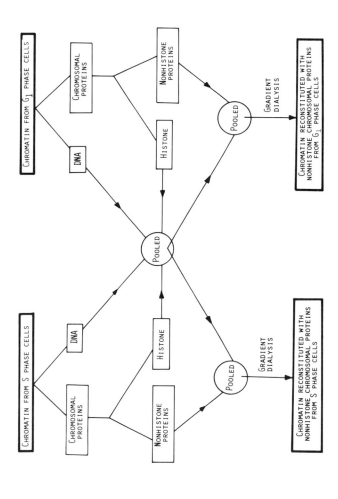

Fig. 9. Flow diagram of experimental protocol for chromatin reconstitution experiment.

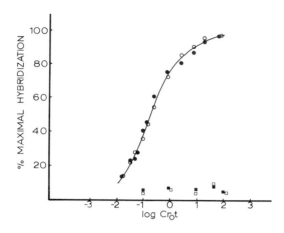

Fig. 10. Kinetics of annealing of histone cDNA to in vitro transcripts from native and reconstituted chromatin. 0.37 ng of ^{3}H--cDNA was annealed at 52° with either 0.15 μg or 1.5 μg of RNA transcripts from chromatin reconstituted with S phase nonhistone chromosomal proteins (o o o o), chromatin reconstituted with G_1 nonhistone chromosomal proteins (■ ■), native S phase chromatin (● ● ● ●) and native G_1 chromatin (□ □). E. coli RNA was included in each reaction mixture so that the final amount of RNA was 3.75 μg.

mosomal proteins regulate the transcription of genes which are transiently expressed.

Conclusions

While it is evident that amongst the nonhistone chromosomal proteins are macromolecules which regulate the readout of specific gene loci, the regulatory elements involved have yet to be defined. Since the genetic sequences for histone messenger RNAs are reiterated (45), the possibility arises that a single cell may contain multiple copies of the proteins which control the expression of histone genes. This system may therefore be readily amenable to the isolation of such regulatory proteins.

Another important question concerns the mechanisms by which nonhistone chromosomal proteins interact with other genome components to render specific sequences transcribable. Nonhistone chromosomal protein phosphorylation has been implicated in this regard (1,4,7,46). Of particular relevance to the regulation of transcription during the cell cycle is the fact that changes in phosphorylation of nonhistone chromosomal proteins have been correlated with cell cycle stage-specific modifications in gene expression (13,14,47). Recently, in collaboration with Lewis Kleinsmith, we (G.S. & J.S.) have obtained direct evidence that the phosphate groups of the nonhistone chromosomal proteins are involved in determining the availability of genes for transcription. These studies are described in a subsequent chapter of this volume.

These studies were supported by research grants from the National Science Foundation (GB 38349), the National Institutes of Health (GM 20535) and the American Cancer Society (F73UF-6 and F75UF-4).

References

1. G. S. Stein, T. C. Spelsberg and L. J. Kleinsmith. Science 183:817 (1974).

2. R. Baserga. Life Sciences 15:1057 (1974).

3. J. Paul and R. S. Gilmour. J. Mol. Biol. 34:305 (1968).

4. C. Teng, C. Teng and V. G. Allfrey. J. Biol. Chem. 246: 3597 (1971).

5. N. Kostraba and T. Y. Wang. Exp. Cell Res. 80:291 (1973).

6. T. C. Spelsberg and L. S. Hnilica. Biochem. J. 120:435 (1970).

7. L. J. Kleinsmith, J. Heidema and A. Carroll. Nature 226: 1025 (1970).

8. G. S. Stein and J. Farber. Proc. Nat. Acad. Sci. 69:2918 (1972).

9. G. S. Stein, S. C. Chaudhuri and R. Baserga. J. Biol. Chem. 247:3918 (1972).

10. G. S. Stein and T. W. Borun. J. Cell Biol. 52:292 (1972).

11. T. W. Borun and G. S. Stein. J. Cell Biol. 52:308 (1972).

12. G. S. Stein and D. Matthews. Science 181:71 (1973).

13. R. Platz, G. S. Stein and L. J. Kleinsmith. Biochem. Biophys. Res. Commun. 51:735 (1973).

14. J. Karn, E. M. Johnson, G. Vidali and V. G. Allfrey. J. Biol. Chem. 249:667 (1974).

15. J. Bhorjee and T. Pederson. Proc. Nat. Acad. Sci. 69: 3345 (1972).

16. E. Gerner and R. Humphrey. Biochim. Biophys. Acta 331:117 (1973).

17. S. Lee, J. Mendecki and G. Brawerman. Proc. Nat. Acad. Sci. 68:1331 (1971).

18. H. Aviv and P. Leder. Proc. Nat. Acad. Sci. 69:1408 (1972).

19. D. Gallwitz and M. Breindl. Biochem. Biophys. Res. Commun. 47:1106 (1972).

20. B. Roberts and B. Paterson. Proc. Nat. Acad. Sci. 70: 2330 (1973).

21. R. Mans and N. Huff. J. Biol. Chem. 250:3672 (1975).

22. R. Ruprecht, N. Goodman and S. Spiegelman. Biochim. Biophys. Acta 294:192 (1973).

23. V. Vogt. Eur. J. Biochem. 33:192 (1973).

24. F. Labrie. Nature 221:1217 (1969).

25. B. D. Young, P. R. Harrison, R. S. Gilmour, G. D. Birnie, A. Hell, S. Humphries and J. Paul. J. Mol. Biol. 84:555 (1974).

26. M. L. Birnstiel, B. H. Sells and I. F. Purdom. J. Mol. Biol. 63:21 (1972).

27. N. A. Straus and T. I. Bonner. Biochim. Biophys. Acta 277:87 (1972).

28. J. O. Bishop. Biochem. J. 113:805 (1969).

29. J. O. Bishop. Biochem. J. 126:171 (1972).

30. T. Barrett, D. Maryanka, P. H. Hamlyn and H. J. Gould. Proc. Nat. Acad. Sci. 71:5057 (1974).

31. S. C. Gulati, D. L. Kacian and S. Spiegelman. Proc. Nat. Acad. Sci. 71:1035 (1974).

32. M. Adesnik and J. Darnell. J. Mol. Biol. 67:397 (1972).

33. C. L. Thrall, W. D. Park, W. H. Rashba, J. L. Stein, R. J. Mans and G. S. Stein. Biochem. Biophys. Res. Commun. 61: 1443 (1974).

34. J. Spalding, K. Kajiwara and G. Mueller. Proc. Nat. Acad. Sci. 56:1535 (1966).

35. E. Robbins and T. W. Borun. Proc. Nat. Acad. Sci. 57:409 (1967).

36. T. W. Borun, M. D. Scharff and E. Robbins. Proc. Nat. Acad. Sci. 58:1977 (1967).

37. D. Gallwitz and G. C. Mueller. J. Biol. Chem. 244:5947 (1969).

38. M. Jacobs-Lorena, C. Baglioni and T. W. Borun. Proc. Nat. Acad. Sci. 69:2095 (1972).

39. T. W. Borun, F. Gabrielli, K. Ajiro, A. Zweidler and C. Baglioni. Cell 4:59 (1975).

40. M. Farquhar and B. McCarthy. Biochem. Biophys. Res. Commun. 53:515 (1973).

41. A. Skoultchi and P. R. Gross. Proc. Nat. Acad. Sci. 70: 2840 (1973).

42. G. S. Stein, W. D. Park, C. L. Thrall, R. J. Mans and J. L. Stein. Biochem. Biophys. Res. Commun. 63:945 (1975).

43. J. L. Stein, C. L. Thrall, W. D. Park, R. J. Mans and G. S. Stein. Science, in press (1975).

44. J. Paul, R. S. Gilmour, N. Affara, G. Birnie, P. Harrison, A. Hell, S. Humphries, J. Windass and B. Young. Cold Spring Harbor Symp. 38:885 (1973).

45. M. C. Wilson, M. Melli and M. L. Birnstiel. Biochem. Biophys. Res. Commun. 61:404 (1974).

46. L. J. Kleinsmith. J. Cell. Physiol., in press (1975).

47. D. Pumo, G. S. Stein and L. J. Kleinsmith, manuscript submitted.

48. G. S. Stein, R. J. Mans, E. J. Gabbay, J. L. Stein, J. Davis and P. D. Adawadkar. Biochemistry 14:1859 (1975).

49. G. S. Stein, W. D. Park, C. L. Thrall, R. J. Mans and J. L. Stein. Manuscript submitted.

THE IN VITRO TRANSCRIPTION OF THE GLOBIN GENE IN CHROMATIN

R. Stewart Gilmour and John Paul
The Beatson Institute for Cancer Research
Glasgow, Scotland

Abstract

Chromatin from erythropoietic tissues is transcribed in vitro and the RNA transcripts hybridized to globin cDNA to assay for the amounts of globin mRNA sequences present. One of the difficulties with this assay is to distinguish endogenous sequences already present in the chromatin from sequences newly synthesized by the bacterial polymerase. Several methods for doing this have been devised. The only practical approach to examining the control elements in chromatin is by transcribing templates that have been reconstituted from isolated components. Again, endogenous RNA can complicate the result. However, it has proved possible to reconstitute chromatin from components previously purified free of endogenous RNA using a CsCl fractionation. These techniques have been applied to mouse foetal liver chromatin and evidence is provided that the control of transcription of the globin gene is through a component of the nonhistone fraction of the chromatin.

An important approach to elucidating the mechanism of gene action in eukaryotes is provided by the study of RNA transcription from isolated chromatin. The discovery that reverse transcriptase from avian myeloblastosis virus can synthesize complementary DNA copies (cDNA) of globin mRNA (1-3) offers a hybridization method for analysing the RNA transcribed in vitro from chromatin for a specific mRNA sequence. Recently several laboratories (4-8) have reported that following the in vitro transcription of a variety of erythroid chromatins globin mRNA sequences can be detected by hybridization to globin cDNA.

We have carried out experiments to demonstrate the transcription by E. coli polymerase of the globin gene in native chromatin from haemopoietic mouse tissues. In contrast to some of the above reports we find that isolated chromatin contains endogenous globin mRNA sequences which can contaminate the in vitro transcripts and give rise to anomalous results.

In this paper it is shown by double labeling and kinetic techniques that globin mRNA sequences are produced de novo as a result of polymerase action.

It has also been shown that haemopoietic (mouse foetal liver) chromatin can be dissociated in 2 M NaCl containing 4 M urea and then reassociated by gradient dialysis to 0.14 M NaCl without affecting the capacity of the chromatin to act as template for globin mRNA sequences. Furthermore there is evidence to suggest that non-erythroid chromatin can prime for globin mRNA following reconstitution in the presence of non-histone proteins prepared by hydroxyapatite chromatography of foetal liver chromatin (11). In this procedure endogenous RNA tends to co-fractionate with nonhistone proteins; therefore hybridization analysis of reconstituted chromatin transcripts is open to the same criticism as those from native chromatin. Here we describe a CsCl fractionation method which yields nonhistone proteins practically devoid of endogenous RNA. Chromatin reconstituted from DNA, histones and CsCl purified nonhistones and transcribed with E. coli RNA polymerase clearly produces de novo globin mRNA sequences and provides a more rigorous proof that the expression of globin genes requires a specific nonhistone protein fraction from erythropoietic tissue.

MATERIALS AND METHODS

Preparation of cDNA.

Globin mRNA was purified from the reticulocytes of mice made anaemic by treatment with phenylhydrazine by passing polysomal RNA twice through a poly(U)-sepharose column. The RNA ran as a single 9S component on polyacrylamide gels and directed the synthesis of mouse globin in the duck lysate cell-free system.

A cDNA copy of globin mRNA was prepared with RNA-dependent DNA polymerase from avian myeloblastosis virus as described by Harrison et al. (15). [^{3}H]dCTP was incorporated into cDNA to give a final specific activity of 20×10^6 dpm/µg.

Estimation of Globin Sequences in RNA Preparations.

In this study we have used a titration technique in which a fixed amount of cDNA (1.0 ng) is annealed at 43°C with increasing amounts of RNA in 10 µl hybridization buffer (0.5 M NaCl; 25 mM HEPES; 1 mM EDTA; 50% formamide pH 6.7). The theoretical considerations of this approach and the treatment of results are described elsewhere by Young et al. (16). Hybrid was measured by the sensitivity of cDNA to single-stranded nuclease

(S_1) prepared from Takadiastase by the method of Sutton (17). The method is demonstrated in Fig. 1 which shows the titration of pure 9S globin mRNA and reticulocyte polysomal RNA to 1 ng

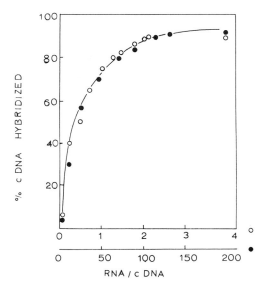

Fig. 1. Titration to 1 ng cDNA of pure 9S globin mRNA (O — O) and reticulocyte polysomal RNA (● — ●).

globin cDNA. Complete hybridization of the pure globin mRNA is achieved at input ratios of RNA:cDNA = 1.6:1, consistent with the finding that cDNA represents a 60% copy of the mRNA. Polysomal RNA required 50X this input to achieve saturation and thus gives an estimate of 2% for globin mRNA content. A small percentage of cDNA (<5%) is normally resistant to S_1 nuclease in the absence of RNA while an additional 10-15% does not appear to be homologous to 9S RNA and is degraded.

Preparation of Chromatin.

Chromatin was prepared by homogenizing tissues in 20 volumes of 1 mM Tris HCl (pH 8), 5 mM $MgCl_2$, and 0.05 mM dithiothreitol. After 10 min, 0.25 volume of 1 M sucrose was added, and the nuclei were sedimented at 2000 x g for 10 min. This step was repeated twice more. The resulting pellet was suspended by homogenization in 20 volumes of 2.2 M sucrose, 1 mM $MgCl_2$ and 0.1 mM dithiothreitol and centrifuged at 30,000 x g for 60 min. The nuclear pellet was washed successively with 10 volumes homogenizing buffer containing 1% Triton X-100 and 10 volumes 0.14 M NaCl, 0.01 M EDTA pH 7; the mixture was sedimented at 2,000 x g for 10 min on each occasion. Finally the pellets were washed with cold distilled water until a clear chromatin gel was obtained.

Preparation of Chromatin Dependent RNA in vitro.

RNA was transcribed in vitro with E. coli RNA polymerase prepared according to the method of Burgess (18).

Incubations (2 ml) contained 0.04 M Tris, pH 7.9; 2.5 mM $MnCl_2$; 1 mM dithiothreitol (DTT); 0.1 mM EDTA; 0.8 mM each ATP, GTP, CTP and UTP; 100 Burgess units RNA polymerase and 500 μg chromatin. After 90 min incubation at 37°C the incubations were cooled. 0.1 ml 1 M $MgCl_2$, 0.4 ml 2 M KCl and 100 μg Worthington DNase were added and the reaction incubated at 37°C for 30 min. Pronase (30 μl of a 5 mg/ml preincubated solution) was added and the incubation continued for 30 min. 1.5 ml 0.2 M EDTA, pH 7.4, 0.2 ml 10% SDS and 3.7 ml phenol (saturated with 0.5 M KCl) were added, mixed thoroughly and heated at 60°C x 5 min. 3.7 ml chloroform:iso-amyl alcohol (99:1) were added, the mixture heated at 60°C for 5 min and centrifuged at 2,000g x 10 min. The organic phase was removed and the remainder re-extracted with chloroform:iso-amyl alcohol at 60°C. The aqueous phase was separated by centrifugation and passed through a Sephadex G50 column in distilled water. Excluded material was collected and dried down.

Reconstitution of Chromatin.

Reconstitution experiments were carried out essentially as described previously (19). Chromatin was dissociated in 2 M NaCl; 5 M urea; 0.01 M Tris HCl, pH 8.3 at a concentration of 0.5 mg DNA/ml and reassociated by dialysing against 5 M urea; 0.01 M Tris HCl, pH 8.3, containing 0.6 M, 0.4 M and 0.2 M NaCl for 16h, 2h and 2h respectively. The resulting precipitate was pelleted and washed with distilled water until a gel formed. As described in the results section, a modified procedure was adopted for the reconstitution with chromosomal proteins purified on CsCl gradients.

RESULTS

Transcription of Mouse Foetal Liver Chromatin.

The foetal liver of the mouse embryo reaches maximum haemopoietic activity after 14 days in utero, at which time about 70% of the cells are erythroid. Chromatin was prepared from 14 day foetal livers and adult mouse brains and incubated with E. coli polymerase. RNA was isolated from a total extraction of the incubation. In a control incubation, mouse foetal liver

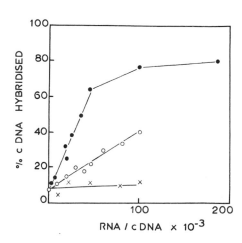

Fig. 2. Titration of RNA transcribed from chromatin with E. coli polymerase, against globin cDNA. Mouse foetal liver chromatin, (●—●); mouse foetal liver chromatin incubated without polymerase, (o—o); mouse brain chromatin, (x——x). In incubations where no polymerase was present E. coli tRNA was added and aliquots hybridized equal to those for complete incubations.

chromatin was incubated without polymerase and an appropriate amount of E. coli soluble RNA added. Fig. 2 shows the results obtained when the purified RNAs were hybridized to 1 ng of globin cDNA.

RNA transcribed from foetal liver chromatin hybridizes to an extent which suggests that 1 part in 6×10^5 of the RNA represents globin mRNA sequences. RNA transcribed from brain fails to hybridize. The contribution due to endogenous RNA in the foetal liver chromatin was estimated by hybridizing identical aliquots of the control and test incubations. Control incubations contained appreciable amounts of hybridizing RNA; however the corresponding values obtained with RNA synthesized in the presence of polymerase were much higher, suggesting a net increase in globin mRNA sequences arising by de novo synthesis.

In the absence of polymerase, however, no RNA is synthesized (as judged by ^{3}H-UTP incorporation) and it is concluded that the hybridization observed in control incubations is due to endogenous globin mRNA sequences present in the chromatin. Foetal liver chromatin contains about 10% by weight of endogenous RNA. Figure 3 shows that isolated endogenous RNA hybridizes to cDNA. About 0.01-0.02% of this RNA represents globin mRNA sequences.

In an attempt to distinguish endogenous globin sequences from those arising de novo, foetal liver chromatin was incubated in the presence and absence of E. coli polymerase for 0,

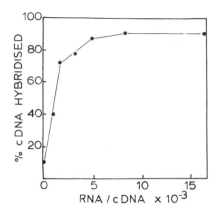

Fig. 3. Titration against cDNA of endogenous RNA prepared from mouse foetal liver chromatin (500 μg).

Fig. 4. Kinetic analysis of globin mRNA content of incubations containing foetal liver chromatin in the presence (●——●) and absence (o——o) of E. coli polymerase. Absolute amounts of globin mRNA were calculated for incubations of 0, 10, 20, 30 and 40 min duration by comparison with a standard 9S titration (Fig. 1).

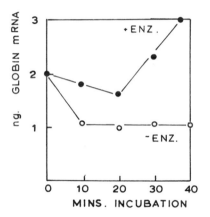

10, 20, 30 and 40 minutes. Where no polymerase was present E. coli soluble RNA was added after half the incubation time had lapsed. The isolated RNAs were hybridized to globin cDNA (Fig. 4). By comparing the titration curves with a standard of pure globin mRNA estimates were made of the total globin mRNA content for each incubation. It is apparent that in the absence of synthesis the net endogenous RNA level falls. Initially the same is true for incubations containing polymerase; however after 30 min there is a net increase in globin sequences over the zero time level.

A more reliable control has been employed for measuring both endogenous RNA and newly synthesized RNA simultaneously in the same incubation. RNA was transcribed from foetal liver chromatin as before, incorporating highly labeled [^{32}P]ATP(1-2 Ci/m mole). After hybridizing to [^{3}H]cDNA, the reaction mix-

ture was diluted to 0.5 ml with 0.2 M NaCl; 0.05 M Tris-HCl
(pH 7.5) and treated with 20 µg/ml pancreatic and T_1 ribonu-
cleases at 30°C for 2 hours. The digest was passed through
Sephadex G50 in 0.2 M NaCl. The excluded material which contained less than 0.1% of the original ^{32}P counts in RNA was
centrifuged to equilibrium in either NaI or CsCl gradients to
separate hybridized cDNA from unhybridized cDNA and [^{32}P]RNA.
In both cases unhybridized [^{32}P]RNA sediments to the bottom of
the gradient. In order to remove as much of this RNA as possible the latter 20% of the gradient was discarded and the remainder containing the [^3H]cDNA made up to original volume and
re-run. Fig. 5 shows the distribution of [^3H] and [^{32}P] ob-

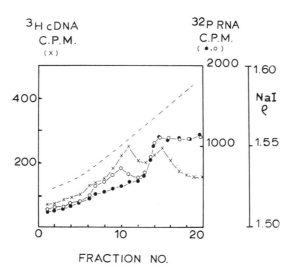

FRACTION NO.

Fig. 5. Isopycnic banding in NaI of the hybrid formed between [^3H]cDNA and [^{32}P]RNA transcribed from mouse foetal liver
chromatin. Counts in [^3H] (x——x); counts in ^{32}P, (o——o);
density (----). A sample of [^{32}P]RNA treated in an identical
fashion but without hybridization to cDNA was run in a separate gradient, (●——●).

tained from the NaI gradient run according to the method of
Birnie et al (12). [^3H] counts appeared at densities of 1.5550
and 1.5720 corresponding to RNA/DNA hybrid and unhybridized
cDNA. [^{32}P] counts were associated with the unhybridized cDNA
in amounts from which it could be calculated that [^{32}P]RNA comprised about 40% of the hybridized RNA. In a parallel gradient
[^{32}P]RNA was subjected to the same procedure but without hybridization to cDNA. Here no peak of [^{32}P]RNA was found in

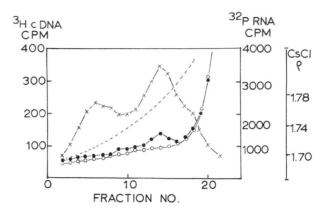

Fig. 6. Isopycnic banding in CsCl of the hybrid formed between [^{3}H]cDNA and [^{32}P]RNA transcribed from mouse foetal liver chromatin. Counts in [^{3}H], (x——x); counts in [^{32}P], (•——•); density (dashed line). A sample of [^{32}P]RNA treated in an identical fashion but without hybridization to cDNA was run in a separate gradient, (o——o).

the hybrid region.

In another experiment the [^{32}P]RNA and [^{3}H]cDNA was analysed on CsCl gradients as described by Szybalski (13). As can be seen in Fig. 6 [^{3}H] counts appeared at densities of 1.78 corresponding to RNA/DNA hybrid and 1.71 corresponding to unhybridized cDNA. From the [^{32}P] counts associated with the hybridized cDNA it was calculated that [^{32}P]RNA accounted for 60% of the hybridized RNA.

Reconstitution Experiments.

Reconstitution experiments were devised to test whether the specific transcription of the globin gene in haemopoietic chromatin is an inherent property of the chromatin and if so which structural elements of the chromatin are responsible for directing this specificity. The functional integrity of foetal liver chromatin was investigated by dissolving the chromatin in 2 M NaCl; 5 M urea; 0.01 M Tris HCl, pH 8.3. This procedure dissociates the DNA and protein components of chromatin. The mixture was reconstituted by gradient dialysis as described in Materials and Methods.

RNA transcribed from reconstituted chromatin was compared with that from native chromatin by hybridization to globin

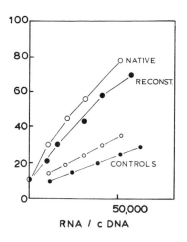

Fig. 7. Titration against globin cDNA of RNA transcribed from
native mouse foetal liver chromatin, (0——0); reconstituted
mouse foetal liver chromatin,
(●——●); RNA from incubations
containing native, (o——o), and
reconstituted chromatins (●——●)
and no polymerase.

cDNA. Control incubations contained chromatin but no polymerase. Fig. 7 shows that RNA transcribed from both chromatins
hybridized to about the same extent; however it can be seen
from control incubations that the reconstitution process does
not eliminate the background due to endogenous RNA.

In further experiments dissociated foetal liver chromatin
was fractionated on hydroxyapatite columns according to the
method of MacGillivray et al (11) to give a nonhistone preparation essentially free of histone and DNA. Chromatin was
prepared from L5178Y cells (a non-haemopoietic cell line) and
dissociated in 2 M NaCl; 5 M urea; 0.01 M Tris HCl, pH 8. A
sample of foetal liver nonhistone proteins was added to the
dissociated L5178Y chromatin (corresponding to half the DNA
concentration). The mixture was reconstituted and the chromatin re-isolated. RNA was synthesized from L5178Y chromatin
reconstituted in the presence and absence of foetal liver nonhistone proteins and compared by titration against globin cDNA
(Fig. 8). In the absence of nonhistone proteins, RNA from
L5178Y chromatin does not hybridize to globin cDNA; however
when reconstituted in the presence of acidic proteins, this
chromatin now supports the transcription of globin mRNA sequences. It is evident from a control containing the chromatin
reconstituted with nonhistone proteins and incubated in the
absence of polymerase that some endogenous RNA is carried
through with the foetal liver proteins.

Two types of experiment have been carried out to ensure
that the presence of endogenous RNA in the reconstitution is
not producing spurious results. Firstly, a double label exper-

27

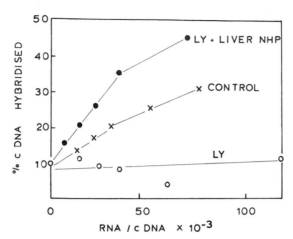

Fig. 8. Titration against globin cDNA of RNA transcribed from L5178Y chromatin reconstituted alone, (0——0) and in the presence of nonhistone proteins from mouse foetal liver, (●——●). RNA was isolated from the latter incubation when no polymerase was present, (x——x).

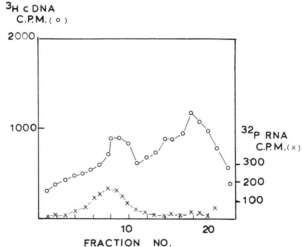

Fig. 9.
Isopycnic band-
ing in CsCl of the hybrid formed between [^{3}H] and [^{32}P]RNA transcribed from LY chromatin reconstituted in the presence of nonhistone proteins isolated from mouse foetal liver. Counts in [^{3}H], (o——o); counts in [^{32}P], (x——x) due to hybridized RNA. Background due to [^{32}P] RNA treated identically but without hybridization to cDNA has been subtracted.

iment was carried out as described previously, in which [³H] cDNA was hybridized to [³²P]RNA synthesized in vitro from chromatin reconstituted in the presence of liver nonhistone proteins. Fig. 9 shows the CsCl analysis of radioactivity showing [³²P] counts sedimenting with [³H] counts at a density of 1.78 corresponding to RNA/cDNA hybrid. It was estimated that about 60% of the RNA in the DNA/RNA hybrid resulted from de novo synthesis. In the second experiment chromosomal proteins were prepared by a method that excluded the presence of endogenous RNA. Foetal liver chromatin was prepared as far as the 0.14 M NaCl wash. Pelleted material was dissolved in 55% CsCl; 4 M urea; 0.01 M Tris pH 8; 0.01 M dithiothreitol; 0.01 M EDTA to a concentration of 300 µg DNA/ml. 4-5 ml aliquots of dissolved chromatin were centrifuged in an MSE 10 x 10 ml Titanium rotor (the remainder of the tube was filled with light paraffin) at 40,000 r.p.m. for 40 hours at 8°C. 0.5 ml fractions of the gradient were collected and monitored at 280 nm. The histones and nonhistones are found in the top 1.5 ml of the gradient while the DNA and RNA form a pellet.

The histone/nonhistone preparation was mixed with purified mouse DNA in 2:1 (w/w) proportions and dialysed against 2 M NaCl; 4 M urea; 0.01 M Tris HCl, pH 8; 0.01 M dithiothreitol and 0.01 M EDTA for 2 hours. Gradient dialysis was then carried out for 15 hours against 0.5 M NaCl in the same urea medium, for 2 hours against 0.14 M NaCl plus urea medium, then overnight against 0.14 M NaCl; 0.01 M EDTA alone. The final reconstituted chromatin was centrifuged at 2,000g x 10 min and washed twice in distilled water to form a gel.

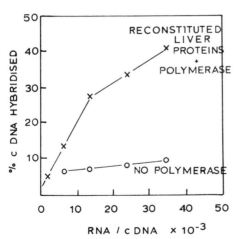

Fig. 10. Titration against globin cDNA of RNA transcribed from mouse foetal liver chromatin reconstituted from purified mouse DNA and CsCl purified chromatin protein in the presence (X——X) and absence (0——0) of E. coli polymerase.

RNA transcribed from foetal liver reconstituted chromatin in the presence of E. coli polymerase was hybridized to 1 ng globin cDNA (Fig. 10). A control incubation without polymerase was carried out as before. The results show that little, if any endogenous RNA is present in the reconstituted chromatin. In the presence of polymerase de novo globin mRNA sequences are transcribed.

This modified reconstitution technique was extended to show that the nonhistone component of foetal liver chromatin is specifically required for the expression of globin genes in the reconstituted chromatin. Chromosomal proteins from both foetal liver and mouse brain were prepared on CsCl/urea gradients. The histones and nonhistones from each were separated by chromatography on hydroxyapatite. The histone fractions were pooled. Samples of nonhistone protein from foetal liver or brain were reconstituted with purified mouse DNA and pooled histones. (In both cases protein:DNA = 2:1.) The chromatins were reconstituted and transcribed. Fig. 11 shows the hybridi-

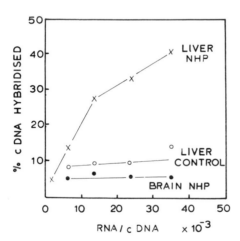

Fig. 11. Titration to cDNA of RNA transcribed from chromatin reconstituted from a common pool of mouse DNA and histones and CsCl purified nonhistone proteins from mouse foetal liver (x——x) and mouse brain (●——●). A control incubation was set up containing reconstituted mouse foetal liver chromatin, E. coli soluble RNA and no polymerase (o——o).

zation to globin cDNA of the isolated RNAs. It can be seen that the globin gene is available for transcription only when foetal liver proteins are present in the reconstitution mixture. Chromatin reconstituted with brain nonhistone proteins fails to hybridize. The background obtained with reconstituted foetal liver chromatin in the absence of polymerase is only slightly above this level.

DISCUSSION

In a previous publication we reported that in vitro transcripts of foetal liver chromatin contained little endogenous
globin sequences (6). Both Axel et al (4) and Steggles et al
(5) reported similar findings with different haemopoietic systems. We have re-examined the question of a suitable control
for chromatin primed reactions in view of the finding that
when carrier RNA is present in control incubations and total
RNA extractions are carried out substantial amounts of endogenous globin mRNA sequences can be detected (Fig. 1). Barrett
et al (8) have also reported the presence of endogenous globin
RNA in chromatin from chick reticulocytes.

The reason for this discrepancy is not clear. It can be
seen from Fig. 4 however, that the amount of endogenous
globin mRNA in control incubations decreases with time. This
could be due to small amounts of ribonuclease activity in the
chromatin. Indeed we have shown that when foetal liver chromatin is pre-incubated for longer periods of time in the absence of polymerase almost all endogenous globin RNA sequences
can be removed (14).

We have attempted to demonstrate the de novo synthesis of
globin mRNA sequences by E. coli polymerase in the presence of
the endogenous sequences in foetal liver chromatin. While it
is possible to show a net increase in the total globin RNA
sequences in the presence of polymerase (Fig. 4), the question
still remains as to whether a "no enzyme" incubation gives a
true estimate of the endogenous contamination in conditions of
RNA synthesis. The use of a double label analysis of hybrids
(Figs. 5, 8, 6) provides the least ambiguous answer. In all
cases it was found that 50% or more of the hybridizing globin
RNA sequences could be attributed to newly transcribed RNA.
These controls lend weight to the conclusion that the chromatins of differentiated haemopoietic tissues have active globin
genes while in other tissues this gene is repressed (Fig. 1).

The functional reconstitution of chromatin offers a method
of investigating the underlying mechanisms of this control.
However it is apparent from the data of Figs. 7 and 8 that
when native foetal liver chromatin is reconstituted or when
a heterologous chromatin is reconstituted in the presence of
nonhistone proteins from foetal liver appreciable amounts of
endogenous globin mRNA sequences are still detectable. Again
double labeling can be used to show that the de novo synthesis
of globin mRNA from reconstituted templates is dependent on
the presence of nonhistone proteins from haemopoietic chromatin.

31

From the point of view of employing reconstitution as a routine assay for the detection of possible gene activators within non-histone fractions, interpretation of the data would be greatly facilitated if RNA-free components could be prepared. It is possible to do this by fractionating chromatin proteins from nucleic acid on 55% CsCl; 4 M urea equilibrium gradients. Fig. 10 shows that foetal liver proteins prepared in this way can be reconstituted to purified mouse DNA to yield a function-ally active chromatin that is essentially devoid of endogenous RNA. Thus it is possible to confirm the conclusions drawn from the data of Fig. 1 in a more decisive fashion. By frac-tionating the CsCl purified nonhistone proteins of foetal liver and brain on hydroxyapatite and reconstituting them in-dividually to a common pool of histones and purified DNA it can be seen (Fig. 11), without recourse to elaborate controls, that the activity of the globin gene is under direct influence of the nonhistone component of the haemopoietic chromatin. In agreement with this finding Barrett et al (8) have recently used a urea/guanidinium chloride centrifugation technique to prepare RNA-free nonhistone proteins from chick reticulocytes, and shown that this fraction can activate the globin genes in reconstituted erythrocyte chromatin.

References

1. J. Ross, H. Aviv, C. Scolnick and P. Leder. Proc. Nat. Acad. Sci. U.S.A. 69:264 (1972).

2. I. M. Verma, G. F. Temple, H. Fan and D. Baltimore. Nature New Biol. 235:163 (1972).

3. D. L. Kacian, S. Spiegelman, A. Bank, M. Terada, S. Meta-phora, L. Dow and P. A. Marks. Nature New Biol. 235:167 (1972).

4. R. Axel, H. Cedar and G. Felsenfeld. Proc. Nat. Acad. Sci. U.S.A. 70:2029 (1973).

5. A. W. Steggles, G. N. Wilson, J. A. Kantor, D. J. Picciano, A. K. Falvey and W. F. Anderson. Proc. Nat. Acad. Sci. U.S.A. 71:1219 (1974).

6. R. S. Gilmour and J. Paul. Proc. Nat. Acad. Sci. U.S.A. 70:3440 (1973).

7. R. S. Gilmour, P. R. Harrison, J. D. Windass, N. A. Affara and J. Paul. Cell Differentiation 3:9 (1974).

8. T. Barrett, D. Maryanka, P. H. Hamlyn and H. J. Gould. Proc. Nat. Acad. Sci. U.S.A. 71:5057 (1974).

9. J. Paul, R. S. Gilmour, N. Affara, G. Birnie, P. Harrison, A. Hell, S. Humphries, J. Windass and B. Young. Cold Spring Harbor Symp. 38:885 (1973).

10. R. S. Gilmour, S. E. Humphries, E. C. Hale and J. Paul. In: "Normal and Pathological Protein Synthesis", INSERM Colloquium, Paris, May 1973, p. 89.

11. A. J. MacGillivray, A. Cameron, R. J. Krauze, D. Rickwood and J. Paul. Biochim. Biophys. Acta 277:384 (1972).

12. G. D. Birnie. FEBS Lett. 27:19 (1972).

13. W. Szybalski. In: "Methods in Enzymology", Vol. XII Part B, L. Grossman and K. Moldave, eds. Academic Press, New York and London (1968), p. 330.

14. R. S. Gilmour, J. D. Windass, N. Affara and J. Paul. J. Cell. Physiol., in press (1975).

15. P. R. Harrison, A. Hell and J. Paul. FEBS Lett. 24:73 (1972).

16. B. D. Young, P. R. Harrison, R. S. Gilmour, G. D. Birnie, Anna Hell, S. Humphries and J. Paul. J. Mol. Biol. 84:555 (1974).

17. W. D. Sutton. Biochim. Biophys. Acta 240:522 (1971).

18. R. R. Burgess. J. Biol. Chem. 244:6160 (1969).

19. R. S. Gilmour and J. Paul. FEBS Lett. 9:242 (1970).

STIMULATION AND INHIBITION OF TRANSCRIPTION IN VITRO BY NONHISTONE CHROMOSOMAL PROTEINS

Tung Yue Wang and Nina C. Kostraba
Division of Cell and Molecular Biology
State University of New York at Buffalo
Buffalo, New York 14214

Abstract

Two nonhistone protein fractions have been isolated from Ehrlich ascites tumor chromatin: one that activates and the other which suppresses transcription of DNA in vitro. The activator nonhistone protein fraction was prepared from the loosely bound nonhistone proteins by phenol extraction and selective binding to tumor DNA, resulting in a phosphoprotein-enriched fraction. When this nonhistone protein fraction is added to a homologous DNA-dependent RNA polymerase II reaction, it stimulated RNA synthesis in vitro. This activation is specific in that it is effective only when homologous DNA and RNA polymerase are used as template and enzyme source. The inhibitor nonhistone protein was isolated from the DNA-protein complex by phenol extraction as an electrophoretically homogeneous protein, with a calculated molecular weight of 12,000 daltons. It binds to the tumor DNA and, when added to a homologous DNA-dependent RNA polymerase reaction, inhibits RNA synthesis in vitro. This nonhistone protein inhibits RNA synthesis by acting at the initiation of the RNA chain, prior to formation of the first phosphodiester bond. These observations suggest that nonhistone proteins are involved in both positive and negative controls of gene activity.

In 1969, with the introduction of the technique of chromatin reconstitution (1,2), Bekhor et al. (1), Huang and Huang (3) and Gilmour and Paul (4) demonstrated that the nonhistone chromosomal proteins are involved in the specific transcription of chromatin in vitro. This was confirmed a year later by Gilmour and Paul (5) and Spelsberg and Hnilica (6). That nonhistone proteins are capable of counteracting inhibition of DNA-dependent RNA synthesis in vitro by histones was observed as early as 1965 by Langan (7,8) and was subsequently supported by similar findings (9,10). These observations suggest a positive role for the nonhistone proteins in the control of gene activity. Direct evidence supporting such a regulatory role

35

for the nonhistone proteins has been provided by Teng, Teng and Allfrey (11,12) and Kleinsmith and associates (13,14) who showed that a phosphoprotein-rich nonhistone protein fraction, which binds specifically to homologous DNA, activates transcription of DNA in vitro. It has also been shown that nonhistone proteins contain a fraction that stimulates transcription of chromatin in vitro (15-18). This activated transcription of chromatin is tissue-specific and the activated transcript codes for different polypeptides than the unstimulated chromatin transcript. Thus, nonhistone proteins contain at least two potential regulatory fractions, both rich in phosphoprotein, operative in positive control of eukaryotic gene expression.

Indirect evidence, however, has suggested that nonhistone proteins are also involved in selective restriction of DNA. For example, dehistonized chromatin (19,20) or sonicated chromatin which exhibits no dissociation of histones (21), still possesses a restricted template capacity in RNA synthesis in vitro. In mitotic cells, RNA synthesis is reduced as compared with S-phased cells. This reduction seems to be caused by nonhistone proteins (22,23).

In this report, we describe two nonhistone protein fractions from Ehrlich ascites tumor chromatin that affect the transcription of DNA in vitro. One fraction, derived from the loosely bound chromatin proteins, activates DNA-dependent RNA synthesis, supporting the original findings of Allfrey and Kleinsmith and their associates (11-14). The other fraction, obtained from the tightly bound chromatin proteins, inhibits transcription of DNA in vitro. Both fractions bind to homologous DNA and contain phosphoproteins. RNA polymerase II, purified from Ehrlich ascites tumor cells, was used as the enzyme source for in vitro RNA synthesis.

Nonhistone Protein Fraction That Activates Transcription of DNA in vitro (24)

The loosely bound nonhistone proteins were isolated from Ehrlich ascites tumor chromatin by extraction with 0.35 M NaCl in 0.02 M Tris-HCl, pH 7.5. The extracted protein was dialyzed against 0.4 M NaCl-0.01 M Tris-HCl, pH 7.0, and passed through a BioRex 70 (Na$^+$) column. The unadsorbed protein was collected, dialyzed against 0.05 M NaCl in 0.01 M Tris-HCl, pH 7.4, containing 0.001 M EDTA, and passed through a DNA-free cellulose column to remove non-specific adsorbing material. The loosely bound nonhistone proteins were then applied to an E. coli DNA-cellulose column, collecting the run-off fraction, which represented the nonhistone proteins that were not bound to heterologous DNA, and were loaded onto an Ehrlich ascites tumor DNA-

cellulose column. The tumor DNA-cellulose column was washed
with 0.05 M NaCl-0.001 M EDTA-0.01 M Tris-HCl, pH 7.4, and
eluted with 0.6 M NaCl-0.001 M EDTA-0.01 M Tris-HCl, pH 7.4.
The 0.6 M NaCl eluted nonhistone protein fraction, representing
the specific homologous DNA binding proteins, was collected
and used for the study of its effect on transcription in vitro.

The specific DNA-binding nonhistone protein fraction obtained as above constituted 1.5% of the total loosely bound
nonhistone proteins, and contained 5% RNA and 0.90% alkali-
labile phosphorus. Based on the phosphoprotein content, the
DNA-binding step resulted in a 30-fold purification from the
total 0.35 M NaCl-soluble nonhistone proteins. The specific
DNA-binding nonhistone protein fraction was still heterogeneous,
having an enhancement of protein species with molecular weights
less than 36,000.

In a homologous DNA-dependent RNA polymerase II reaction,
in which both DNA and the enzyme were prepared from Ehrlich
ascites tumor cells, the addition of the specific DNA-binding
nonhistone protein fraction stimulated RNA synthesis (Fig. 1).

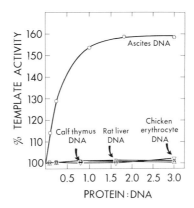

Fig. 1. Stimulation of DNA-dependent RNA synthesis in
vitro by 0.35 M NaCl-soluble DNA-binding nonhistone protein
fraction. The RNA synthesis system was as described in the
legend to Table 1 and contained 5 µg DNA and the DNA-binding
nonhistone protein fraction in amounts as indicated.

This activation of transcription in vitro was not observed when
Ehrlich ascites tumor DNA was substituted by DNA's prepared
from calf thymus, rat liver and chicken red blood cells. Furthermore, when Ehrlich ascites tumor RNA polymerase II was replaced by Micrococcus luteus RNA polymerase, there was no stimulation of RNA synthesis by the nonhistone protein fraction

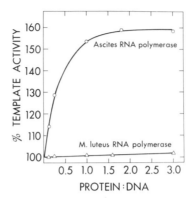

Fig. 2.   Preference for homologous RNA polymerase in the stimulation of DNA-dependent RNA synthesis in vitro by 0.35 M NaCl-soluble DNA-binding nonhistone protein fraction.   Experimental conditions were as in Fig. 1.

(Fig. 2). These results indicate that the loosely bound non-histone proteins contain a phosphoprotein-rich fraction that can activate template- and RNA polymerase-specific RNA synthesis in vitro.

Two points may be made of this study. First, since the specific DNA-binding tumor nonhistone protein fraction had no effect on transcription in the M. luteus RNA polymerase reaction, it may be assumed that the activation by the nonhistone protein fraction is enzyme specific, perhaps effective only with eukaryotic RNA polymerases. Secondly, the nonhistone protein fraction described in the present work was purified from the loosely bound nonhistone proteins, obtained by extraction of chromatin with 0.35 M NaCl. It has been suggested that these loosely bound nonhistone proteins are contaminating cytoplasmic proteins (25,26). Indeed, removal of the loosely bound nonhistone proteins from chromatin by 0.35 M NaCl has been reported to alter neither the structure nor the template activity of the chromatin (19). However, it has also been shown by Baserga and associates that extraction with low salt of chromatin from WI-38 cells stimulated to proliferate depresses the chromatin template activity to that of the unstimulated control (27) and alters chromatin structure as detected by circular dichroism (28). Similar structural alteration of the chromatin by low salt has been reported by Hjelm and Huang (29). These results and the present data indicate that the loosely bound nonhistone proteins are functional chromosomal proteins.

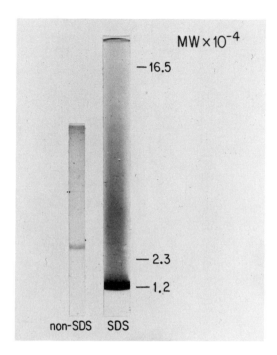

Fig. 3. Polyacrylamide gel electrophoresis of nonhistone protein isolated from Ehrlich ascites tumor DNA-protein complex. Left panel: 5 μg of the nonhistone protein were subjected to electrophoresis on non-denaturing 5% polyacrylamide gel for 5.5 hr at 0.2 mA per gel. Right panel: 20 μg of the nonhistone protein were electrophoresed on 10% SDS-polyacrylamide gel for 24 hr at 0.5 mA per gel. Both gels were stained with Coomassie Brilliant blue.

A Nonhistone Protein That Inhibits Transcription of DNA in vitro

The chromatin was prepared from Ehrlich ascites tumor nuclei, extracted with 2.0 M NaCl in 0.02 M Tris-HCl, pH 8.0, and the extract was dialyzed against 13 volumes of 0.01 M Tris-HCl, pH 8.0. The precipitated DNA-protein complex was collected by centrifugation, and extracted with 0.4 N H_2SO_4 to remove the histones. The acid-extracted residue was isolated by phenol extraction (12). The phenol-soluble nonhistone protein thus isolated appeared as a single electrophoretic protein band either by nondenaturing or by SDS-polyacrylamide gel electrophoresis (Fig. 3). From the latter, the unit molecular weight was calculated to be 12,000. This nonhistone protein contained 2.7% alkali-labile phosphorus, and had an acidic to basic amino acid residues of 1.4. When chromatographed on an Ehrlich ascites tumor DNA-cellulose column, the nonhistone protein was eluted by 1.0 M NaCl. There were no detectable nuclease and protease activities in the protein.

When supplemented to a homologous DNA-dependent RNA polymerase II reaction, the nonhistone protein inhibited RNA synthesis in vitro (Fig. 4). This inhibitory effect of the non-

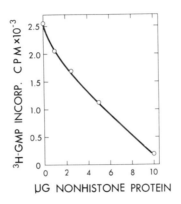

Fig. 4. Inhibition of DNA-dependent RNA synthesis in vitro by the nonhistone protein. The RNA synthesis system was as in Fig. 1 except that the nonhistone protein used was isolated from Ehrlich ascites tumor DNA-protein complex as described in the text.

histone protein was not significantly altered by excess RNA polymerase, but reduced by increasing amount of DNA (Table 1). This result indicates that the inhibitory action of the non-histone protein in the DNA-dependent RNA polymerase reaction is on the template, not on the enzyme.

The data in Table 2 illustrate the mode of action of the nonhistone protein on the DNA-dependent RNA synthesis in vitro. If the nonhistone protein was added to the RNA polymerase reaction mixture consisting of only the initiation complex, or the initiating ribonucleoside triphosphate, and/or an additional ribonucleoside triphosphate, there was no inhibition of RNA synthesis in vitro by the nonhistone protein. However, if the nonhistone protein was introduced to the system prior to the addition of the initiating ribonucleoside triphosphate, there was full inhibition. The data indicate that the nonhistone protein inhibits transcription in vitro by interacting with

TABLE 1

Effect of Template and Enzyme Concentration
on RNA synthesis Inhibited by Nonhistone Protein

DNA (μg)	RNA polymerase (unit)	% inhibition
5	0.05	42
10	0.05	30
20	0.05	21
5	0.01	39
5	0.02	35
5	0.05	39

The reaction mixture, in a total volume of 0.25 ml, contained the following: Tris-HCl, pH 7.9, 10 μmoles; $MnCl_2$, 0.75 μmole; $MgCl_2$, 1.15 μmoles; $(NH_4)_2SO_4$, 12.5 μmoles; EDTA, 0.017 μmole, β-mercaptoethanol, 1.0 μmole; ATP, CTP and UTP, 0.0625 μmole each; ^{3}H-GTP (1 μCi), 0.00625 μmole; nonhistone protein, 3.7 μg; and Ehrlich ascites tumor DNA and RNA polymerase II, as indicated. The reaction mixture was incubated at 37°C for 1 hr, and the radioactivity of the acid-insoluble precipitate was determined as described elsewhere (24).

DNA, preventing the initiation of RNA chain at the step prior to the formation of the first phosphodiester bond.

These studies demonstrate that nonhistone chromosomal proteins contain fractions that bind to homologous DNA and can activate and restrict transcription in vitro. The results suggest that nonhistone proteins are potential regulatory molecules involved in both positive and negative controls of gene expression, resulting in differential transcription in eukaryotic cells.

Acknowledgement

This work was supported in part by a U. S. Public Health Service Grant (GM-11698).

TABLE 2

Inhibition of RNA Synthesis by Nonhistone Protein

	Pre-incubation	Additions	RNA synthesis (pmoles ^{3}H-ATP incorp.)
1.	DNA	(A,C,G,U)TP	29
2.	DNA + NHP	(A,C,G,U)TP	6
3.	DNA + ATP	NHP + CTP + GTP + UTP	26
4.	DNA + ATP + CTP	NHP + GTP + UTP	28
5.	DNA + ATP + GTP	NHP + CTP + UTP	29
6.	DNA + ATP + UTP	NHP + CTP + GTP	28
7.	DNA + (A,C,G,U)TP	NHP	37
8.	DNA + (A,C,G,U)TP	---	39
9.	DNA + NHP + (A,C,G,U)TP	---	4

Pre-incubation of DNA, RNA polymerase and other indicated components was 10 min., followed by further incubation with the additions as shown for 50 min. Where indicated, the amount of nonhistone protein added was 10 μg. The assay mixture was as described in Table 1.

References

1. I. Bekhor, G. M. Kung, and J. Bonner. J. Mol. Biol. 39: 351 (1969).

2. R. C. C. Huang. In: "Recent Development in Biochemistry", H. W. Li et al., eds. Republic of China (cited in R. C. C. Huang and P. C. Huang, J. Mol. Biol., 39:365 (1969)),(1968).

3. R. C. C. Huang and P. C. Huang. J. Mol. Biol. 39:365 (1969).

4. R. S. Gilmour and J. Paul. J. Mol. Biol. 40:137 (1969).

5. R. S. Gilmour and J. Paul. FEBS Lett. 9:242 (1970).

6. T. C. Spelsberg and L. S. Hnilica. Biochem. J. 120:435 (1970).

7. T. A. Langan and L. Smith. Inform. Exch. Group No. 7 (1965).

8. T. A. Langan. In: "The Regulation of Nucleic Acid and Protein Synthesis", V. V. Koningsberger and L. Bosch, eds. Elsevier, Amsterdam (1967), p. 233.

9. T. Y. Wang. Exp. Cell Res. 53:288 (1968).

10. T. Y. Wang. Exp. Cell Res. 57:467 (1969).

11. C. T. Teng, C. S. Teng and V. G. Allfrey. Biochem. Biophys. Res. Commun. 41:690 (1970).

12. C. S. Teng, C. T. Teng and V. G. Allfrey. J. Biol. Chem. 246:3597 (1971).

13. M. Shea and L. J. Kleinsmith. Biochem. Biophys. Res. Commun. 50:473 (1973).

14. L. J. Kleinsmith. J. Biol. Chem. 248:5648 (1973).

15. T. Y. Wang. Exp. Cell Res. 61:455 (1970).

16. M. Kamiyama and T. Y. Wang. Biochim. Biophys. Acta 220: 563 (1971).

17. N. C. Kostraba and T. Y. Wang. Biochim. Biophys. Acta 262:169 (1972).

18. N. C. Kostraba and T. Y. Wang. Cancer Res. 32:2348 (1972).

19. T. C. Spelsburg and L. S. Hnilica. Bicchim. Biophys. Acta 228:202 (1971).

20. J. Paul and R. S. Gilmour. J. Mol. Biol. 34:305 (1968).

21. A. T. Ansevin, L. S. Hnilica, T. C. Spelsberg and S. Kehm. Biochemistry 10:4793 (1971).

22. J. Farber, G. Stein and R. Baserga. Biochem. Biophys. Res. Commun. 47:790 (1972).

23. G. Stein, G. Hunter and L. Lavie. Biochem. J. 139:71 (1974).

24. N. C. Kostraba, R. A. Montagna and T. Y. Wang. J. Biol. Chem. 250:1548 (1975).

25. E. W. Johns and S. Forrester. Eur. J. Biochem. 8:547 (1969).

26. D. E. Commings and L. O. Tack. Exp. Cell Res. 82:175 (1973).

27. L. H. Augenlicht and R. Baserga. Transplant Proc. 3:1177 (1973).

28. J. C. Lin, C. Nicolini and R. Baserga. Biochemistry 13: 4127 (1974).

29. R. P. Hjelm, Jr. and R. C. C. Huang. Biochemistry 13:5275 (1974).

DO PHOSPHORYLATED PROTEINS REGULATE GENE ACTIVITY?

Lewis J. Kleinsmith
Department of Zoology
University of Michigan
Ann Arbor, Michigan

Abstract

Nonhistone chromosomal proteins are phosphorylated and de-phosphorylated within the intact nucleus by two independent sets of reactions, a protein kinase reaction which transfers the terminal phosphate group of a variety of nucleoside and deoxy-nucleoside triphosphates to serine and threonine residues in the proteins, and a phosphatase reaction which cleaves these phosphoserine and phosphothreonine bonds and releases inorganic phosphate. Several lines of evidence are consistent with the hypothesis that the phosphorylation and dephosphorylation of these proteins is involved in gene control mechanisms, including the finding that phosphorylated nonhistone proteins are highly heterogeneous and their phosphorylation patterns are tissue specific, changes in their phosphorylation correlate with changes in chromatin structure and gene activity, and phos-phorylated nonhistone proteins bind specifically to DNA.

Fractionation of nuclear protein kinases demonstrates the existence of at least twelve enzyme fractions capable of phos-phorylating nonhistone proteins whose activities vary in both their substrate specificities and susceptibility to regulation via cyclic AMP. Heterogeneity is also evident in the nuclear protein phosphatases involved in the dephosphorylation of non-histone proteins. A nonhistone protein phosphatase which is devoid of detectable protease activity has been purified from the nuclear sap fraction of calf thymocytes. By covalently linking this phosphatase to agarose, a tool can be created for use in selectively dephosphorylating nonhistone proteins prior to chromatin reconstitution. In this way experiments can be designed for directly answering for the first time the question of whether phosphorylated proteins regulate gene activity.

Although the exact roles played by the various DNA-asso-ciated proteins of higher organisms are not yet known, evidence accumulated over the past several years has led to the conclu-sion that the molecules responsible for regulating the activity of individual genes are to be found in the nonhistone protein fraction. It was the pioneering chromatin reconstitution experiments of Gilmour and Paul (1,2) which first clearly focused attention on this property of nonhistone proteins, and the preceding three papers in this volume represent some of the more recent developments along these lines. Amongst the other properties of nonhistone proteins which support the notion of a genetic regulatory function for at least some of these molecules are their heterogeneity and tissue specificity, the changes in their composition and metabolism occurring during alterations in gene activity, and the specificity of their binding to DNA (for recent reviews see 3,4).

Although such evidence has strongly implicated nonhistone proteins in the process of gene regulation, it has not told us very much about how this control is exerted. There is one special property of nonhistone proteins, however, which I feel may provide an important clue in this regard, and that is the extensive occurrence of phosphorylation and dephosphorylation reactions in this protein fraction. My interest in the phosphorylation of nonhistone proteins dates back to 1965, when Thomas Langan first told me of his discovery of large quantities of the amino acid phosphoserine in nuclear proteins. Since the occurrence of phosphorylated amino acids in proteins is rather unusual (although it was a great deal more unusual in 1965 than it is today!), it seemed reasonable at the time to speculate that the high concentration of phosphate in nuclear proteins has an important functional significance. Over the years we have gradually learned more and more about the reactions involved in the phosphorylation and dephosphorylation of nuclear proteins, and a number of lines of evidence now point to the conclusion that these events are in some way involved in the process of genetic regulation (for recent reviews see 5,6).

In spite of this supportive evidence, direct evidence for a role of phosphorylated proteins in regulating the activity of specific genes is still lacking. In this presentation I would like to first briefly review several of the properties of phosphorylated nonhistone proteins which are compatible with the notion that protein phosphorylation plays a role in gene regulation and then outline some recent experiments which for the first time provide a way for directly testing whether phosphorylated proteins regulate the activity of specific genes.

Phosphorylation and Dephosphorylation of Purified Nonhistone Phosphoproteins

The phosphorylated nonhistone protein fraction which we have focused most of our attention upon is isolated by a salt extraction procedure originally described by Langan (7), and subsequently modified in our laboratory (8,9). In essence this procedure involves washing nuclei with dilute salt solutions to remove soluble proteins of the nuclear sap, solubilizing the chromatin with 1.0 M NaCl, diluting the salt concentration to 0.4 M to precipitate histones and DNA, and selectively purifying the phosphorylated nonhistone proteins from the supernatant by adsorption with calcium phosphate gel. The final protein fraction usually contains about 1.0-1.3% phosphorus by weight, which is sufficient to phosphorylate 4-5 amino acids out of every 100 present. The major phosphorylated amino acid is phosphoserine (roughly 90%), but significant amounts of phospho-threonine can also be detected. The phosphorylated proteins present in this fraction are thus a subclass of the nonhistone proteins which have been selectively purified on the basis of their high phosphate content.

Experiments carried out in the late 1960s (10-12) on the enzymological properties of these phosphoproteins led to the general model summarized in Figure 1. According to this model

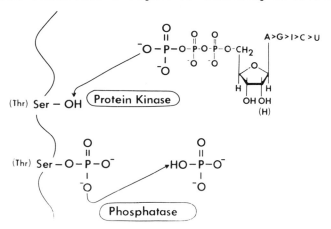

Non-Histone
Phosphoprotein

Fig. 1. Model summarizing the reactions involved in the phosphorylation and dephosphorylation of nonhistone proteins. Serine (and to a much lesser extent, threonine) residues are phosphorylated in a protein kinase reaction involving the ter-minal phosphate group of various nucleoside and deoxynucleoside triphosphates. In a separate reaction phosphate groups already present in the nonhistone proteins are cleaved off by a phos-phatase and released as inorganic phosphate.

47

the nonhistone proteins are phosphorylated via a protein kinase reaction in which the terminal phosphate of nucleoside or deoxynucleoside triphosphates is esterified to a serine (or threonine) residue. Although ATP is one of the best phosphate donors, several other nucleoside and deoxynucleoside triphosphates work quite well. The protein kinase activities which catalyze this reaction are endogenous to the purified nonhistone phosphoprotein fraction. In a second type of reaction the phosphate groups bound to the nonhistone proteins are removed via a phosphatase, which releases them in the form of inorganic phosphate. The fact that this dephosphorylation reaction is distinct from the phosphorylation reaction is demonstrated by the finding that the enzymes responsible for dephosphorylation are not present in the purified phosphoprotein fraction, as are the protein kinases. Such dephosphorylating activity can be shown to be present in the intact nucleus, however.

The finding that nonhistone proteins are continually being phosphorylated at the expense of high-energy phosphate compounds, such as ATP, while the protein-bound phosphate groups are being released in the low energy form of inorganic phosphate indicates that considerable amounts of energy are being consumed in the process. The question therefore arises as to functional significance of these energy-consuming reactions. There is now ample evidence indicating that the addition and removal of phosphate groups is employed to regulate the functional properties of other proteins (13,14), and so the obvious possibility arises that the phosphorylation and dephosphorylation of nonhistone proteins may be involved in controlling the structural and metabolic properties of chromatin. Since one of the key metabolic events occurring in chromatin is gene transcription, we have proposed that nonhistone prosphorylation is involved in the control of gene activity. This hypothesis has been used to generate a number of specific predictions about the behavior of nonhistone phosphoproteins which can be easily tested.

Nonhistone Phosphoproteins are Heterogeneous and Tissue-Specific

If phosphorylated nonhistone proteins are involved in regulating the activity of specific genes, one would predict that these proteins ought to be heterogeneous and exhibit differences among tissues where differences in gene expression occur. Using the technique of SDS-acrylamide gel electrophoresis it has been confirmed that in contrast to the histones, the nonhistone phosphoproteins are in fact highly heterogeneous (15-17). Although the banding patterns exhibit fundamental similarities from tissue to tissue, reproducible differences

TABLE I

Summary of Correlations Reported Between Phosphorylation of
Nonhistone Proteins and Gene Activity

System	Reference
Growth and Development	
Lectin-stimulated lymphocytes	18-20
Avian erythrocytes	21
Physarum polycephalum	22
HeLa cells	23,24
Yoshida sarcoma	25
Sea urchin embryos	26
Muscle differentiation	27
Neuronal & glial cells	28
Kidney regeneration	29
Chinese hamster cells (K_{12})	30
WI-38 fibroblasts	31
Hamster kidney cells ($BHK_{21}C_{13}$)	32
Hormonal Stimulation	
Testosterone (prostate)	33,34
Glucocorticoids (liver)	35-37
Aldosterone (kidney)	38
Estradiol (uterus)	39
Prolactin (mammary gland)	40
Chorionic gonadotropin (ovary)	41
Malignant and Transformed Cells	
Breast carcinoma	42
Azo-dye carcinogenesis	43-45
SV-40 transformed fibroblasts	46-48
Drugs	
Isoproterenol (salivary glands)	49,50
Phenobarbital (liver)	51
α-1,2,3,4,5,6-Hexachlorocyclohexane (liver)	52

can be observed. Such differences are most striking when one
compares the phosphorylation patterns of ^{32}P-labeled nonhistone
proteins obtained from various sources.

Nonhistone Phosphorylation Correlates with Gene Activity

Another prediction which follows logically from the hy-

pothesis that phosphorylation of nonhistone proteins is involved in the regulation of gene transcription is that alterations in the phosphorylation of these proteins ought to correlate with changes in gene activity. This prediction has been
tested in several experimental systems in our laboratory, and
has also been explored by a large number of other investigators.
Since it would not be feasible to describe the data from more
than one or two systems in this presentation, it is perhaps
more impressive to simply summarize all the different types of
systems in which such correlations have been observed. As can
be seen from examining Table I, nonhistone protein phosphorylation has been shown to undergo dynamic alterations during the
normal growth and development of a wide variety of cell types,
during stimulation by steroid and peptide hormones, during
malignant transformation, and as a result of treating cells
with various drugs. Since all these situations involve changes
in gene transcription, a strong correlation between nonhistone
phosphorylation and gene expression is clearly established.
Although these findings support our initial hypothesis, correlation experiments in themselves can not prove the existence of
a cause-and-effect relationship. For this reason we have
turned our attention to other experimental approaches which may
more directly address the question of whether the nonhistone
phosphoproteins are genetic regulatory molecules.

Nonhistone Phosphoproteins Bind Specifically to DNA

Genetic regulatory molecules isolated from bacterial systems have been found to be acidic proteins which recognize and
bind to specific sites on the chromosomal DNA. In order to
determine whether the phosphorylated nonhistone proteins found
in eukaryotic cells bind to DNA in a similar fashion, we have
exploited the technique of DNA-cellulose chromatography (53,54).
At physiological ionic strength 1-2% of the purified nonhistone
phosphoprotein fraction binds to its homologous DNA. With foreign DNAs the binding of these proteins is considerably less,
indicating the specificity of the reaction. The phosphorylated
proteins which exhibit this behavior consist of a heterogeneous
collection of polypeptides when analyzed via SDS-acrylamide gel
electrophoresis, but most of them fall in the molecular weight
range of 30,000-70,000. Specific binding of phosphorylated
nonhistone proteins to DNA has also been demonstrated by others
employing sucrose gradient centrifugation (16).

Cyclic AMP and Nuclear Protein Kinases

If nonhistone phosphorylation is involved in the regulation of gene expression, the phosphorylation reaction would in

turn be expected to be subject to some type of control mechanism. The most obvious candidate for such a regulator of protein phosphorylation is cyclic AMP. Although many studies on nuclear protein kinase activity have failed to demonstrate any effects of this nucleotide, under appropriate conditions the protein kinases associated with the purified nonhistone phosphoprotein fraction can be subfractionated on phosphocellulose columns into 12 distinct enzymatic activities, some of which are stimulated by cyclic AMP and some of which are inhibited by it (9). The earlier failures to observe such effects of cyclic nucleotides can probably be accounted for by the masking of these opposing effects which occurs when analyses are performed on cruder, unfractionated preparations.

Nuclear Protein Phosphatases

Although all the lines of evidence described thus far are consistent with the hypothesis that the phosphorylation of nonhistone proteins is involved in the control of eukaryotic gene transcription, none of them provides direct proof for this concept. One of the most straightforward approaches for obtaining such direct proof would be to determine the effects of nonhistone proteins, before and after dephosphorylation, on the ability of specific genes to be transcribed by RNA polymerase. This approach requires a mechanism for selectively dephosphorylating nonhistone proteins, for which reason Kathy Leonardson working in our laboratory has recently begun attempts to purify nuclear protein phosphatases. An important consideration in attempting to purify such enzymes is the particular assay for protein phosphatase employed. The release of ^{32}P from labeled protein into an acid-soluble form is insufficient because any protease which cleaves off small ^{32}P-containing peptides will be incorrectly detected as a phosphatase. For this reason we have been careful to design an assay based on selective extraction of released inorganic ^{32}P only (Table II).

TABLE II

Outline of Assay Procedure for Nonhistone Protein Phosphatase*

1. Incubate ^{32}P-labeled nonhistone protein with enzyme fraction to be tested
2. Precipitate protein with silicotungstic acid
3. Add ammonium molybdate to form phosphomolybdate complex with released inorganic ^{32}P
4. Extract ^{32}P-labeled phosphomolybdate complex with 1/1 iso-butanol/benzene
5. Measure radioactivity in organic phase

*For details of procedure see refs. 10-12.

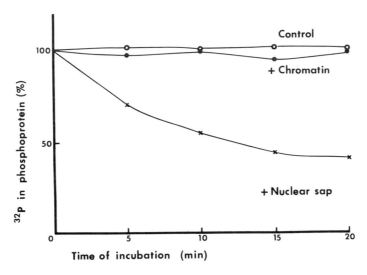

Fig. 2. Caption on following page.

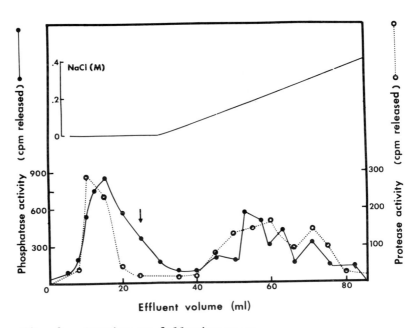

Fig. 3. Caption on following page.

Using this assay with ^{32}P-labeled nonhistone proteins as a substrate, we have not found significant levels of nonhistone phosphatase activity to be present in chromatin or any of the chromosomal protein fractions. We have, however, observed considerable activity in the nuclear sap fraction (Figure 2). Chromatography of this material on DEAE-Sephadex columns demonstrates the existence of multiple protein phosphatases within this fraction (Figure 3). Unfortunately, most regions where phosphatase activity elute also contain protease activity. This clearly makes them unsuitable as sources of enzyme for selectively dephosphorylating nonhistone proteins, since the treated proteins would be at least partially degraded in the process.

One region eluted from the DEAE-Sephadex column, however, does contain appreciable nonhistone phosphatase activity which is completely free of demonstratable proteolytic activity (see arrow in Figure 3). This material has therefore been chosen as a tool for use in dephosphorylating nonhistone proteins. By mixing the enzyme fraction with cyanogen bromide-activated agarose, it can be linked to an insoluble matrix. In this way the enzyme can be easily removed after the nonhistone proteins are dephosphorylated. In the following paper it will be shown how this material can be used in conjunction with chromatin reconstitution to make possible experiments which for the first time are capable of directly answering the question of whether phosphorylated proteins are involved in the regulation of the transcription of specific genes.

Fig. 2 (above). Release of inorganic phosphate from ^{32}P-labeled purified nonhistone phosphoprotein catalyzed by various nuclear fractions. Preparation of labeled phosphoprotein and conditions of assay are described elsewhere (11,12). Note that the nuclear sap contains an active protein phosphatase.

Fig. 3 (below). Column chromatography of nuclear sap (0.14 M NaCl extract of calf thymus nuclei) on DEAE-Sephadex run in 0.01 M Tris-HCl (pH 8.1) and eluted with increasing concentrations of NaCl. Nonhistone phosphatase and protease activities were measured in each fraction. Note that there are multiple peaks of phosphatase activity, but only in one region (arrow) is the phosphatase free of detectable protease activity. This material was collected for use in the experiments described in the following paper.

Acknowledgments

These studies were supported in part by grants GB 23921 and BMS 74-23418 from the National Science Foundation. A portion of these experiments were performed while the author was the recipient of a John Simon Guggenheim Memorial Foundation Fellowship.

References

1.   J. Paul and R. S. Gilmour. J. Mol. Biol., 34:305 (1968).

2.   R. S. Gilmour and J. Paul. FEBS Lett., 9:242 (1970).

3.   G. S. Stein, T. C. Spelsberg and L. J. Kleinsmith. Science 183:817 (1974).

4.   I. L. Cameron and J. R. Jeter, Jr., eds. "Acidic Proteins of the Nucleus", Academic Press, New York (1974).

5.   L. J. Kleinsmith. In: "Acidic Proteins of the Nucleus", I. L. Cameron and J. R. Jeter, Jr. eds. Academic Press, New York (1974).

6.   L. J. Kleinsmith. J. Cell.Physiol. 85:459 (1975).

7.   T. A. Langan. In: "Regulation of Nucleic Acid and Protein Biosynthesis", V. V. Koningsberger and L. Bosch eds. Elsevier, Amsterdam (1967), p. 103.

8.   E. L. Gershey and L. J. Kleinsmith. Biochim. Biophys. Acta 194:331 (1969).

9.   V. M. Kish and L. J. Kleinsmith. J. Biol. Chem. 249:750 (1974).

10.   L. J. Kleinsmith, V. G. Allfrey and A. E. Mirsky. Proc. Nat. Acad. Sci. USA 55:1182 (1966).

11.   L. J. Kleinsmith and V. G. Allfrey. Biochim. Biophys. Acta 175:136 (1969).

12.   L. J. Kleinsmith and V. G. Allfrey. Biochim. Biophys. Acta 175:123 (1969).

13.   E. G. Krebs. Curr. Topics Cell Regul. 5:133 (1972).

14.   H. L. Segal. Science 180:25 (1973).

15. R. D. Platz, V. M. Kish and L. J. Kleinsmith. FEBS Lett.
     12:38 (1970).

16. C. S. Teng, C. T. Teng and V. G. Allfrey. J. Biol. Chem.
     246:3597 (1971).

17. D. Rickwood, P. G. Riches and A. J. MacGillivray.
     Biochim. Biophys. Acta 299:162 (1973).

18. L. J. Kleinsmith, V. G. Allfrey and A. E. Mirsky. Science
     154:780 (1966).

19. B. G. T. Pogo and J. R. Katz. Differentiation 2:119
     (1974).

20. E. M. Johnson, J. Karn and V. G. Allfrey. J. Biol. Chem.
     249:4990 (1974).

21. E. L. Gershey and L. J. Kleinsmith. Biochim. Biophys.
     Acta 194:519 (1969).

22. W. M. LeStourgeon and H. P. Rusch. Science 174:1233
     (1971).

23. R. D. Platz, G. S. Stein and L. J. Kleinsmith. Biochem.
     Biophys. Res. Commun. 51:735 (1973).

24. J. Karn, E. M. Johnson, G. Vidali and V. G. Allfrey.
     J. Biol. Chem. 249:667 (1974).

25. P. G. Riches, K. R. Harrap, S. M. Sellwood, D. Rickwood
     and A. J. MacGillivray. Biochem. Soc. Trans. 1:70 (1973).

26. R. D. Platz and L. S. Hnilica. Biochem. Biophys. Res.
     Commun. 54:222 (1973).

27. N. T. Mân, G. E. Morris and R. J. Cole. FEBS Lett. 42:
     257 (1974).

28. H. Fleischer-Lambropoulos, H.-I. Sarkander and W. P.
     Brade. FEBS Lett. 45:329 (1974).

29. W. P. Brade, J. A. Thomson, J.-F. Chiu and L. S. Hnilica.
     Exp. Cell Res. 84:183 (1974).

30. M. Rieber and J. Bacalao. Exp. Cell Res. 85:334 (1974).

31. D. Pumo, G. S. Stein and L. J. Kleinsmith, in preparation.

32. M. Marty de Morales, C. Blat and L. Harel. Exp. Cell Res. 86:111 (1974).

33. K. Ahmed and H. Ishida. Mol. Pharmacol. 7:323 (1971).

34. P. Schauder, B. J. Starman and R. H. Williams. Proc. Soc. Exp. Biol. Med. 145:331 (1974).

35. V. G. Allfrey, E. M. Johnson, J. Karn and G. Vidali. In: "Protein Phosphorylation in Control Mechanisms", F. Huijing and E. Y. C. Lee, eds. Academic Press, New York, (1973), p. 217.

36. G. D. Bottoms and R. A. Jungmann. Proc. Soc. Exp. Biol. Med. 144:83 (1973).

37. P. Schauder, B. Starman and R. H. Williams. Experientia 30:1277 (1974).

38. C. C. Liew, D. Suria and A. G. Gornall. Endocrinology 93: 1025 (1973).

39. M. E. Cohen and L. J. Kleinsmith. Fed. Proc. 34:704 (1975).

40. R. W. Turkington and M. Riddle. J. Biol. Chem. 244:6040 (1969).

41. R. A. Jungmann and J. S. Schweppe. J. Biol. Chem. 247: 5535 (1972).

42. N. Kadohama and R. W. Turkington. Cancer Res. 33:1194 (1973).

43. J. -F. Chiu, C. Craddock, S. Gatz and L. S. Hnilica. FEBS Lett. 33:247 (1973).

44. K. Ahmed. Res. Commun. Chem. Pathol. Pharmacol. 9:771 (1974).

45. J. -F. Chiu, W. P. Brade, J. Thomson, Y. H. Tsai and L. S. Hnilica. Exp. Cell Res. 91:200 (1975).

46. M. O. Krause, L. J. Kleinsmith and G. S. Stein. Exp. Cell Res. (1975), in press.

47. M. O. Krause, L. J. Kleinsmith and G. S. Stein. Life Sci. 16:1047 (1975).

48. D. Pumo, G. S. Stein and L. J. Kleinsmith. Biochim. Biophys. Acta, in press.

49. H. Ishida and K. Ahmed. Exp. Cell Res. 78:31 (1975).

50. H. Ishida and K. Ahmed. Exp. Cell Res. 84:127 (1974).

51. J. Blankenship and E. Bresnick. Biochim. Biophys. Acta 340:218 (1974).

52. W. P. Brade, J. -F. Chiu and L. S. Hnilica. Mol. Pharmacol. 10:398 (1974).

53. L. J. Kleinsmith, J. Heidema and A. Carroll. Nature 226: 1025 (1970).

54. L. J. Kleinsmith. J. Biol. Chem. 248:5648 (1973).

DIRECT EVIDENCE FOR A FUNCTIONAL
RELATIONSHIP BETWEEN NONHISTONE CHROMOSOMAL
PROTEIN PHOSPHORYLATION AND GENE TRANSCRIPTION

Lewis J. Kleinsmith*, Janet Stein** and Gary Stein**
*Department of Zoology
University of Michigan
Ann Arbor, Michigan

and

**Department of Biochemistry
University of Florida
Gainesville, Florida

Abstract

     In order to examine the question of whether the phosphory-
lation state of nonhistone proteins is involved in determining
their ability to regulate gene transcription, a purified nonhistone protein phosphatase linked to agarose was employed to
selectively dephosphorylate these proteins prior to chromatin
reconstitution. Using proteins and DNA obtained from HeLa
chromatin prepared from synchronized S-phase cells, it was
found that dephosphorylation of the nonhistone proteins results
in approximately a 50% reduction in the number of template
sites available for the initiation of transcription. Furthermore, specific measurements of histone gene transcription using
a complementary DNA probe show that dephosphorylation of nonhistone proteins specifically inhibits the ability of these
genes to be transcribed. Taken together these results directly
implicate protein phosphorylation as a factor involved in regulating the transcription of individual genes in eukaryotes.

     The phosphorylation of nonhistone chromosomal proteins has
been implicated as playing a potentially important role in determining the availability of genetic sequences for transcription in many biological systems (1-3). Over the past several
years our laboratories have been collaborating on studies aimed
at elucidating the involvement of nonhistone protein phosphorylation in the regulation of transcription during the cell cycle.
The rationale for these studies has been that while direct evidence from chromatin reconstitution experiments indicates that

nonhistone proteins are responsible for cell cycle stage-specific differences in chromatin template activity (4) and mediate the transient readout of histone genes (Stein et al., this
volume), the manner in which these proteins render genetic
sequences available for transcription remains unresolved. The
addition and removal of phosphate groups may provide a viable
mechanism for such regulation. We have observed that there
are pronounced variations in the synthesis and turnover of
nonhistone phosphoproteins during G_1, S, G_2 and mitosis in con-
tinuously dividing cells (5,6; also see 7), and that changes in
the metabolism of nonhistone phosphoproteins occur following
stimulation of contact inhibited WI-38 human diploid fibroblasts to proliferate (8). Although these results are of a
correlative nature and hence circumstantial, they are consistent with the hypothesis that nonhistone protein phosphorylation is involved in the activation and repression of genes during the cell cycle. To examine more directly the possibility
that the phosphorylation of nonhistone proteins is functionally
related to transcriptional properties of the genome within this
context, a series of chromatin reconstitution experiments have
been carried out in which the nonhistone proteins were dephosphorylated prior to reconstitution.

Effects of Nonhistone Dephosphorylation on Chromatin Transcription

Chromatin from synchronous HeLa cells in S-phase was dissociated in 3 M NaCl-5 M urea, the DNA removed by centrifugation, and the histones and nonhistones separated on QAE-Sepha-
dex as described elsewhere (4). Prior to chromatin reconstitution a portion of the nonhistone proteins was partially dephosphorylated by passing it over a column of agarose to which
the protein phosphatase described in the preceding chapter had
been covalently attached. Since the columns must be run in
5 M urea to prevent precipitation of nonhistone proteins, the
enzymatic activity is not optimal, but 30 to 50% dephosphorylation can be achieved in this manner. Another batch of nonhistone proteins was run over a plain agarose column to serve as
control. SDS-acrylamide gel staining patterns of these control
and dephosphorylated nonhistone proteins were found to be in-
distinguishable. Chromatin was then reconstituted by combining
these nonhistone proteins with histones and DNA, and subjecting
the mixture to gradient dialysis (4). The overall design of
this experimental approach is summarized in Figure 1.

Measurements of the template activity of reconstituted
chromatins with added E. coli RNA polymerase showed about a 50%
reduction in chromatin reconstituted in the presence of par-

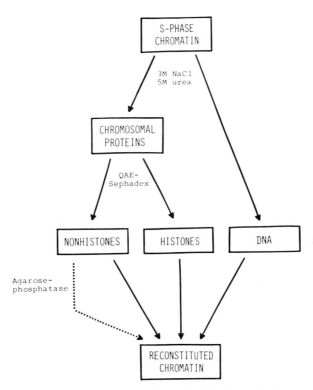

Fig. 1.   Diagram summarizing the design of the experiments on reconstituted chromatin.

tially dephosphorylated nonhistone proteins (Figure 2). Because routine template activity assays can not distinguish between effects on initiation, elongation, or termination, direct measurements of the number of initiation sites were also performed (9).   In this approach initiation is allowed to proceed in the absence of CTP, which limits the extent to which elongation can occur to the point where the first C is required.   High salt is then added to block further initiation, and the previously initiated sites are allowed to elongate via adding CTP. In this way each available initiation site is transcribed into one chain of RNA.   By keeping the RNA polymerase concentration low and measuring the amount of chromatin required to use up all the RNA polymerase molecules, one can determine the number of initiation sites available.   As is shown in Figure 3, chromatin reconstituted in the presence of partially dephosphorylated nonhistone proteins has roughly 50% less initiation sites than control reconstituted chromatin.

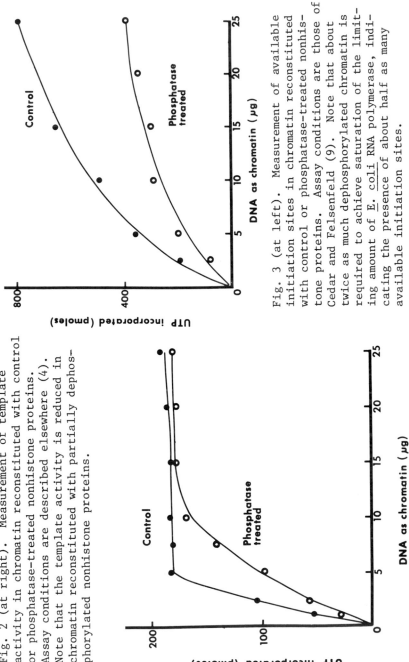

Fig. 2 (at right). Measurement of template activity in chromatin reconstituted with control or phosphatase-treated nonhistone proteins. Assay conditions are described elsewhere (4). Note that the template activity is reduced in chromatin reconstituted with partially dephosphorylated nonhistone proteins.

Fig. 3 (at left). Measurement of available initiation sites in chromatin reconstituted with control or phosphatase-treated nonhistone proteins. Assay conditions are those of Cedar and Felsenfeld (9). Note that about twice as much dephosphorylated chromatin is required to achieve saturation of the limiting amount of E. coli RNA polymerase, indicating the presence of about half as many available initiation sites.

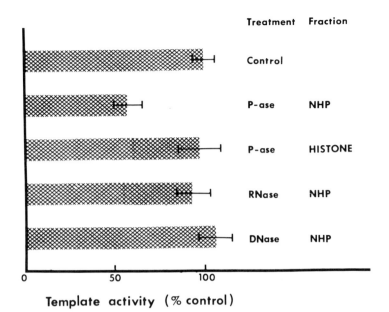

	Treatment	Fraction
	Control	
	P-ase	NHP
	P-ase	HISTONE
	RNase	NHP
	DNase	NHP

Template activity (% control)

Fig. 4. Effects of various treatments of chromosomal proteins prior to reconstitution on the template activity of the
subsequently reconstituted chromatin. All enzymes were insol-
ubilized by covalent attachment to cyanogen bromide-activated
agarose, in order to permit their removal prior to chromatin
reconstitution. P-ase = phosphatase (collected from region of
arrow in Figure 3 of previous paper), NHP = nonhistone protein.
Note that only dephosphorylation of nonhistone proteins affects
template activity in this system.

In order to determine whether these effects are specifically related to the dephosphorylation of nonhistone proteins,
a number of other control experiments have been performed. As
is shown in Figure 4 neither dephosphorylation of histones, nor
RNase or DNase treatment of the nonhistones, produces such a
drop in chromatin template activity.

Effect of Nonhistone Dephosphorylation on Transcription of Histone Genes

Although the results presented thus far implicate the phosphorylation state of nonhistone proteins as a key factor in
determining the overall number of initiation sites available
for transcription, one can not distinguish whether this effect
is random or specifically directed at individual sites. In

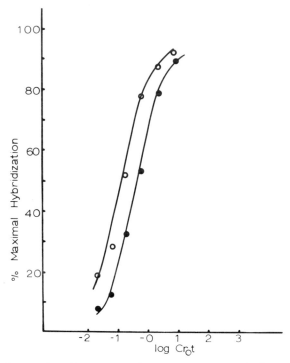

Fig. 5. Hybridization of RNA transcribed from reconstituted chromatin to histone gene complementary DNA probe (see Stein et al., this volume, for details). Note that the RNA transcribed from chromatin reconstituted with partially dephosphorylated nonhistone proteins (● ● ● ●) hybridizes at significantly higher $C_{r_0}t$ values than control reconstituted chromatin (o o o), indicating a selective reduction in the availability of histone gene sequences for transcription.

order to approach this issue more directly experiments have been performed in which the transcription of histone genes in reconstituted chromatin was specifically measured using a histone complementary DNA probe (10,11; and Stein et al., this volume). As is shown in Figure 5, the $C_{r_0}t$ curve of hybridization of the transcribed RNA to the complementary histone DNA is shifted to higher values with RNA obtained from chromatin reconstituted in the presence of partially dephosphorylated nonhistone proteins. This shift corresponds to about an 80% reduction in the availability of histone genes for transcription. It should be emphasized that this reduction can not be explained by the 50% decrease in overall chromatin template activity measured previously, because these hybridization analyses measure the relative amount of histone messenger RNA contained within the total population of RNA molecules transcribed from the chromatin template. Thus if the transcription of all gene sequences were randomly inhibited by 50%, the $C_{r_0}t$ curve for histone specific RNA would be identical to controls, because the relative percentage of histone-specific RNA would be unchanged. The fact that the $C_{r_0}t$ curve is displaced after nonhistone dephosphorylation therefore indicates that the effect is somewhat selective for transcription of the histone genes.

Taken together these recent experiments provide the first direct evidence implicating the state of phosphorylation of nonhistone proteins as a controlling factor in determining the availability of individual genes for transcription.

Acknowledgements

These studies were supported in part by grants BMS74-23418, BMS75-18583 and GB 38349 from the National Science Foundation. The experiments were performed while Lewis Kleinsmith was on sabbatical leave in the laboratory of Gary and Janet Stein, Department of Biochemistry, University of Florida Medical School. The support of a John Simon Guggenheim Memorial Foundation Fellowship which made this leave possible is gratefully acknowledged.

References

1. L. J. Kleinsmith. In: "Acidic Proteins of the Nucleus", I. L. Cameron and J. R. Jeter, Jr., eds. Academic Press, New York (1974), p. 103.

2. G. S. Stein, T. C. Spelsberg and L. J. Kleinsmith. Science 183:817 (1974).

3. L. J. Kleinsmith. J. Cell.Physiol. 85:459 (1975).

4. G. Stein and J. Farber. Proc. Nat. Acad. Sci., USA 69: 2918 (1972).

5. R. D. Platz, G. S. Stein and L. J. Kleinsmith. Biochem. Biophys. Res. Commun. 51:735 (1973).

6. D. Pumo, G. S. Stein and L. J. Kleinsmith, in preparation.

7. J. Karn, E. M. Johnson, G. Vidali and V. G. Allfrey. J. Biol. Chem. 249:667 (1974).

8.   D. Pumo, G. S. Stein and L. J. Kleinsmith, in preparation.

9.   H. Cedar and G. Felsenfeld.   J. Mol. Biol. 77:237 (1973).

10.   C. L. Thrall, W. D. Park, H. W. Rashba, J. L. Stein, R. J. Mans and G. S. Stein.   Biochem. Biophys. Res. Commun. 61: 1443 (1974).

11.   G. S. Stein, W. D. Park, C. L. Thrall, R. J. Mans and J. L. Stein.   Biochem. Biophys. Res. Commun. 63:945 (1975).

CHANGES IN NUCLEAR PROTEINS DURING EMBRYONIC DEVELOPMENT AND CELLULAR DIFFERENTIATION

Robert D. Platz, Sidney R. Grimes, Gwen Hord,
Marvin L. Meistrich,* and Lubomir S. Hnilica
University of Texas System Cancer Center
M.D. Anderson Hospital and Tumor Institute
Department of Biochemistry
and
*Department of Experimental Radiotherapy
Houston, Texas

Abstract

The sea urchin embryo and the rat testis are two developmental systems which are well suited for studying the regulatory function of nuclear proteins during differentiation. The sea urchin embryo represents a dynamic system characterized by increasing macromolecular heterogeneity as the cells diverge during cytodifferentiation. In contrast, spermatogenesis in the rat testis represents the differentiation of a single cell type characterized by progressive functional and compositional simplification as the sperm cell matures. In sea urchin embryos, we have focused our attention on the modification of a class of nonhistone proteins which are enzymatically phosphorylated. The rate and specificity of nonhistone protein phosphorylation appears to reflect the process of differentiation thereby indicating that nuclear phosphoproteins may be involved in genetic regulation. Analysis of nuclear proteins from several stages of spermatogenesis (using cells separated by velocity sedimentation) shows a progressive replacement of histone and nonhistone proteins by very basic, low molecular weight proteins. While most nuclear proteins are actively synthesized in the early stages of spermatogenesis, only three low molecular weight, basic proteins are actively synthesized in late stage spermatids. Thus, the synthesis and heterogeneity of chromosomal proteins qualitatively reflects the particular stages of sperm cell differentiation.

The development of a multicellular organism begins with a zygote which contains the total genetic information encoded in its DNA. During the process of differentiation, different cell types arise. Even though each cell contains the same informational DNA, only a part of the available information is expressed. This selective utilization of information contained in DNA occurs during the normal function of mature cells (e.g. in hormone activation, regeneration, etc.) as well as during the establishment of different cell types in the developing embryo. Whether or not the same regulatory mechanisms operate in both situations has not been determined. In both cases however, the question is the same: What are the regulatory mechanisms for determining which genes are transcribed into RNA and which are not?

Transcriptional control mechanisms are frequently conceptualized in terms of repressor-derepressor systems similar to the model of gene regulation proposed for bacteria by Monod and Jacob (1). The simplicity and beauty of this model has insured its persistence in spite of the paucity of experimental evidence to show its operation in eukaryotic cells. On the other hand, the model of gene regulation in higher cells proposed by Britten and Davidson (2) has, for the most part, been consistent with the experimental data from eukaryotic systems. In addition to models for generalized gene regulation, several unique control mechanisms have been described in developmental systems. These include selective chromosomal diminution in Ascaris (3) and inactivation of the X-chromosome in mammals (4). Although these unique controls appear to occur in exceptional situations and represent relatively permanent changes, they provide an indication of the variety of alternative control mechanisms available to, and used by, differentiating cells.

In a broad definition of differentiation, no distinction is made between degrees or types of cell change so long as they are relatively permanent or are considered to contribute to some further change (5). From this definition, it is obvious that differentiation includes cellular changes regulated at the translational as well as the transcriptional level. Consequently, in examining the genetic regulation of differentiation, it is necessary to demonstrate that the changes observed in the cell are the result of regulatory events occuring at the level of RNA synthesis.

That differential gene action is at the basis of cellular differentiation has been deduced from the evidence for gene-controlled enzyme synthesis together with the presence of different arrays of enzymes in different types of cells. However, the most direct evidence of differential gene activation in cellular differentiation is the occurence of differing populations of RNA molecules in various cell types. The requirement for qualitative and quantitative analysis of the RNA products synthesized in specific cell types is being met by utilizing mRNAs for specific proteins to construct complementary DNA molecules (6). The cDNA can then be used to measure small amounts of specific RNA molecules by hybridization (7). If a direct relationship exists between the RNA synthesized and specific cellular changes, the system becomes potentially useful for the identification of regulatory molecules and the in vitro analysis of their mechanisms of action.

Over the past ten years, the search for regulatory molecules has pointed increasingly to the nonhistone nuclear proteins as the class of macromolecules most likely to include proteins functional in the regulation of specific genes. These data have been reviewed recently by Stein, Spelsberg and Kleinsmith (8). Specifically, the nonhistone proteins are present in increased amounts in active tissues and chromatins; they are highly heterogeneous and exhibit tissue and species specificity; they can stimulate the synthesis of RNA in cell-free systems; some of them bind specifically to DNA; and the synthesis of particular classes of these proteins is associated with the induction of gene activity.

We have been interested in the role of nuclear proteins during the differentiation of cells in two systems: spermatogenesis in the rat testis, and early development in the sea urchin embryo. The sea urchin embryo represents a dynamic system characterized by increasing macromolecular heterogeneity as the cells diverge during differentiation (9). Spermatogenesis on the other hand represents a uni-directional and highly ordered sequence of changes characterized by a progressive simplification of macromolecular composition, and by a profound inactivation of RNA synthesis in the maturing spermatid (10). Thus, in the sea urchin embryo we have a rapidly expanding system in terms of the activation of new genes and establishment of new cell types, while in spermatogenesis the trend is toward inactivation of gene transcription and functional specialization.

Spermatogenesis in the Rat Testis.

Spermatogenesis involves a progressive differentiation of the cells of the seminiferous epithelium toward a terminal stage represented by the mature spermatozoan (11). The process begins with a period of proliferation in which spermatogonial cells divide several times to augment the number of potential germ cells. A few cells remain to renew the stem cell population, while most enter a period of growth as primary spermatocytes. The prophase of the first meiotic division is extended, lasting about eight days in the rat. During this time, dramatic changes take place in the nucleus as homologous chromosomes pair and chromosomal rearrangements occur. The first meiotic division is followed immediately by the second, giving rise to haploid spermatids. During the subsequent period of spermatid maturation, the cell is morphologically transformed from a round cell with a large nucleus into a tight package of DNA having a specific morphology and very little cytoplasm. Thus the mammalian testis provides a sequence of well characterized changes in cellular morphology which is readily available in the adult animal. These changes are under hormonal control, but the overall rate at which they take place is highly resistant to modification by external or internal factors.

The feasibility of using the mammalian testis as a source of differentiating cells for biochemical studies has been greatly enhanced by the development of techniques for the separation of testis cells into relatively homogeneous populations (12). One of the techniques we have used employs velocity sedimentation at unit gravity to separate cells according to their size (Figure 1). By applying this technique to rat testis cell

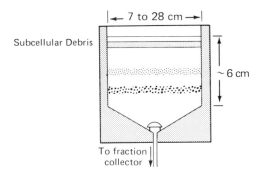

Fig. 1. Diagram of the cylindrical chamber or "Staput" apparatus used to separate cells by velocity sedimentation at unit gravity. The Staput is loaded from the bottom with the cell suspension and followed by a nonlinear 1-4% BSA gradient (42). After 3 1/2 hours, the cells have moved to form discrete bands according to their size, and the chamber is unloaded into a fraction collector. Peak fractions are pooled and concentrated by centrifugation. Nuclei were isolated as described elsewhere (15).

suspensions, we have obtained fractions enriched in specific cell types as shown in Table I. Fractions enriched in pachytene spermatocytes (B), round spermatids (E and F), and late

TABLE I

| | | | | Staput Fractions | | | | |
	B	C	D	E	F	G	H	I
Average Sedimentation rate (mm of migration/hour)	14.0	11.2	9.1	7.0	5.3	3.8	2.4	1.0
Cell Type								
Spermatogonia	1.5	1.2	2	1.6	1.0	8	1.1	0.2
Spermatocytes								
Early Primary[a]	4	4	5	4	9	6	0.8	0.2
Pachytene	/61/	/44/	19	2	0.6	0	0	0
Secondary	0.4	1.2	6	1.4	0.2	0	0	0
Spermatids								
Steps 1–8	13	24	43	/71/	/67/	9	0	0.2
Steps 9–10	2	1.8	2	5	3	6	3	0.2
Steps 11–15	7	10	7	5	7	/15/	/26/	21
Steps 16–19	0.4	1.2	0.4	0.7	0.4	2	4	/27/
Unknown and degenerating	3	4	5	2	3	11	6	0.8
Non-Germinal[b]	3	4	2	2	3	5	0.4	0
Residual Bodies[c]	5	6	8	5	6	38	59	50

Footnotes and caption on next page.

spermatids (H and I) were regarded as representative of those specific cell types or stages of spermatogenesis. This assumption was verified by experiments in which nuclei isolated from the designated fractions were further purified by a second Staput separation (13). This procedure permitted us to obtain fractions containing between 80 and 90 percent pachytene or round spermatid nuclei. Since late stage spermatids are resistant to sonication, they may be obtained at >95% purity simply by sonication of whole testes (14).

Acid extraction of nuclei from the various Staput fractions removes most of the histone and some nonhistone proteins. The electrophoretic patterns of the acid soluble proteins from fractions representing various cell types are clearly different (Figure 2). For example, the spermatidal proteins (TP and S1) which are major components of fractions enriched in late spermatids (G + H and I) are virtually absent from fractions enriched in pachytene spermatocytes (B + C) and round spermatids (D and F). The amount of TP and S1 combined correlates closely with the presence of late stage spermatids in Staput fractions (15). Another example of cell type-specific basic proteins is evident from the ratio of the two lysine-rich histones F1 and X1 (Figure 2), which changes significantly from pachytene spermatocytes (B + C) to early spermatids (F). The high level of X1 in pachytene cells is consistent with the hypothesis that this histone fraction may be involved in the processes of chromosomal pairing and rearrangement which occur during meiosis (16).

Table II – continued below

a	Includes leptotene, zygotene and early pachytene spermatocytes.
b	Includes Sertoli cells, Leydig cells and macrophages. Erythrocytes were not counted.
c	Residual bodies are defined as cytoplasmic fragments which contain ribonucleoprotein aggregates (44). Other cytoplasmic fragments were not scored.

The percentages of rat testis cell types were determined in fractions obtained after 3 hours of sedimentation at unit gravity. The counts were made on random areas of air-dried smears fixed in Bouins and stained with PAS-hematoxylin (43). At least 500 cells were counted in each fraction. The predominant cell type in each fraction is enclosed in a box.

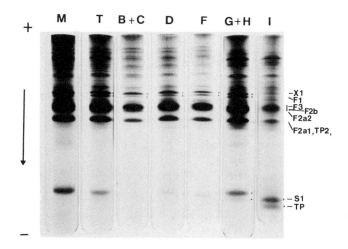

Fig. 2 (above).   See legend on next page.

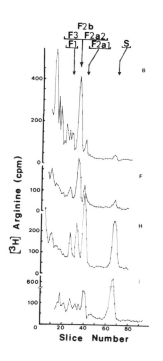

Fig. 3 (at left). Radioactivity profiles of acid soluble proteins from Staput fractions after separation on acid urea polyacrylamide gels (45). Rats were sacrificed 24 hours after intratesticular injection of [^{3}H]arginine (100 μCi/testis). Cylindrical gels were cut into 1 mm slices using a Gilson Aliquogel fractionator and counted by liquid scintillation spectrometry. Profiles are shown from Staput fractions B, F, H and I. Positions of major histones are indicated by arrows; S, spermatidal proteins TP and S1 combined.

The incorporation of [^{3}H]arginine into the acid soluble proteins (Figure 3) indicates a differential synthesis of histones in different Staput fractions. In fraction B, enriched in pachytene spermatocytes, incorporation is predominantly in the region of histones F3, F2b, and F2a2, while in fraction H, enriched in elongating spermatids, incorporation is predominantly in the region of histone F2a1. A significant level of incorporation is consistently observed in the acid soluble nonhistone proteins of fraction B, but we have not yet characterized these proteins. In Fraction I, enriched in later stage spermatids, most of the incorporation occurs in the spermatidal proteins TP and S1. After further purification of late spermatids to greater than 95% purity, the incorporation profile shows two regions of active synthesis in late stage spermatids (Figure 4). These regions of the gel include the spermatidal proteins, TP and S1, and the F2a1-like protein hereafter called TP2.

The identification of bands labeled TP and S1 is based on amino acid analysis of the two major proteins isolated from sonication resistant sperm heads (Table II). Allowing for slight contamination by small amounts of other proteins, the amino acid compositions obtained for these proteins are reasonably close to those reported by Kistler et al. (17). Kistler's

---

Fig. 2. Electrophoretic patterns of acid soluble nuclear proteins from different Staput fractions. Nuclei were isolated from the various cell types, and the basic proteins extracted with 0.25 N HCl were separated by acrylamide gel electrophoresis in 2.5 M urea, pH 2.7 (45). Basic proteins from two controls and five Staput cell fractions are shown.
(M) Acid soluble nuclear proteins from a mechanical preparation of rat testis cells.
(T) Control sample from cells prepared with trypsin but not separated by the Staput.
(B + C) Staput fraction containing primarily late pachytene primary spermatocytes (61%) and round spermatids, steps 1-8 (14%).
(D) Staput fraction containing primary spermatocytes (19%) plus round spermatids (51%), but few late spermatids, steps 9-19 (9%).
(F) Staput fraction containing early round spermatids (56%) and late spermatids (27%).
(G + H) Staput fraction containing late spermatids (33%) and residual bodies (48%).
(I) Staput fraction containing late spermatids (82%) and very small residual bodies (16%).

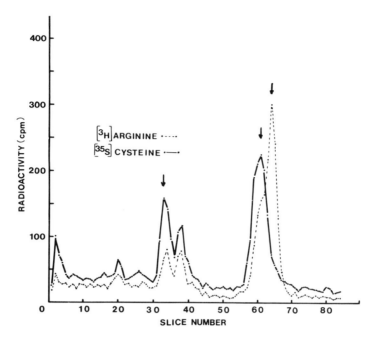

Fig. 4. Profile of radioactivity in acid soluble proteins isolated from sonication resistant spermatids. L-[³H]arginine (100 µCi/testis) and L-[³⁵S]cystine (250 µCi/testis) were administered by intratesticular injection in a total volume of 100 µl. Acid soluble proteins were separated by gel electrophoresis (45), sliced at 1 mm intervals and counted. The peak in slice number 38 is a variable component and has not been identified. TP2 is located in slice number 33. The [³⁵S] peak in slice 61 identifies the position of S1 in relationship to TP in slice 64.

analysis indicated that the slower migrating S1 protein contains 65% arginine and 9.4% cysteine, while the faster migrating TP contains about 20% arginine and 20% lysine, but has no cysteine. The other protein being synthesized in late stage spermatids (TP2) migrates slightly faster than F2al on long gels and contains cysteine. The amino acid composition of this protein band together with that of calf thymus histone F2al is shown in Table III.

TABLE II

Amino Acid	Moles/100 Moles of amino acids recovered[a]	
	TP	S1
Lysine	15.6	5.6
Histidine	4.2	0.4
Arginine	25.3	56.8
Aspartic Acid	6.1	1.2
Threonine	3.0	1.6
Serine	12.6	8.4
Glutamic Acid	1.0	0.7
Proline	1.9	Trace
Glycine	8.9	0.5
Alanine	3.7	1.8
Cysteine[b]	3.2	13.2
Valine	1.6	Trace
Methionine	2.7	Trace
Isoleucine	0.6	Trace
Leucine	4.7	0.5
Tyrosine	4.0	6.5
Phenylalanine	0.5	2.0

[a] No correction was made for hydrolysis of serine or threonine.

[b] Cysteine determined as cysteic acid after performic acid oxidation (47).

Rat testis spermatids were isolated by sonication and centrifugation through 1.5 M sucrose. The heads were reduced and dissolved in 5 M guanidine-HCl, 0.28 β-mercaptoethanol, 0.5 M Tris, pH 8.5 (46). The basic proteins extracted in 0.25 N HCl were fractionated on a Sephadex G-75 column equilibrated with 0.01 N HCl. The final peak fraction containing the rapidly migrating spermatidal proteins was dialyzed against distilled water, lyophilized, and extracted with 0.4 N H_2SO_4. The protein insoluble in H_2SO_4 migrates as S1 (Figure 2) on acrylamide gels. The proteins soluble in H_2SO_4 were fractionated with TCA and the material insoluble between 3 and 20% TCA was identified as TP (Figure 2). These two fractions were hydrolyzed in 6 N HCl for 24 hours at 110°C and analyzed on a modified Beckman Model 120 C amino acid analyzer.

TABLE III

Amino Acid	Moles/100 Moles of amino acids recovered		
	Calf Thymus F2a1[a]	Rat Testis F2a1[b]	TP2[b]
Lysine	10.9	8.9	8.7
Histidine	1.9	1.9	(5.2)
Arginine	13.9	13.1	14.2
Aspartic Acid	5.0	6.2	5.0
Threonine	6.6	7.7	7.9
Serine	2.5	3.0	22.2
Glutamic Acid	6.2	7.8	6.6
Proline	1.3	1.9	13.4
Glycine	15.9	15.9	6.1
Alanine	7.5	7.9	4.5
Valine	7.8	8.7	2.3
Methionine	1.0	0.3	(0.9)
Isoleucine	5.6	5.9	0.3
Leucine	8.0	9.0	2.1
Tyrosine	3.5	0.05	(Trace)
Phenylalanine	2.2	2.2	0.8

[a] Literature value (27).

[b] No correction was made for hydrolysis of serine or threonine.

The testis specific spermatidal protein TP2 was isolated by extraction of purified spermatids in 0.4 N H_2SO_4. The fraction insoluble between 5 and 20% TCA was greatly enriched in TP and TP2. After separation on acrylamide gels (45), the stained bands were cut out and prepared for amino acid analysis as recommended by Houston (48), except that β-mercaptoethanol was not added.

The rat testis F2a1 histone was obtained by electrophoresis of acid soluble protein from whole rat testis nuclei and analyzed in the same way. The amino acid composition of calf thymus F2a1 is included for comparison in order to evaluate the reliability of the procedure. Values for histidine, methionine and tyrosine are not reliable.

The differential pattern of histone synthesis in different cell types suggests that these proteins may be involved in a variety of functions during spermatogenesis. For example, the peak of [^{3}H]arginine incorporation in pachytene cells (Figure 3,B) corresponds in migration to the position of histones F3, F2b, and F2a2. The histones from rat testis cells which migrate in this region have been shown to include additional basic proteins designated X2 and X3 (18). Since these proteins have been found only in meiotic tissues, they may be structural components of the chromosomes with a specific function during meiosis, or they may serve as the pairing proteins which comprise the synaptonemal complex between homologous chromosomes in zygotene stage spermatocytes (19).

The functional significance of the spermatidal proteins is also unclear. They have generally been assigned a role in nuclear condensation and gene inactivation which takes place during spermatid maturation (20). Evidence from kinetic studies (14) in addition to cell separation indicates that TP is synthesized between steps 9 and 15, with a half-life of about 2 days, and S1 is synthesized between steps 15 and 18. The spermatidal protein TP2 is synthesized and degraded with essentially the same kinetics as TP. These testis-specific proteins appear to be transient components during spermatid maturation and may function in the replacement of somatic histones by the sperm protein S1. The formation of disulfide linkages between S1 molecules during sperm maturation in the epididymis renders the cell resistant to disruption by a variety of mechanical and chemical treatments (21). The sperm heads are soluble, however, in guanidine HCl in the presence of β-mercaptoethanol (22).

The synthesis of spermatidal proteins by late spermatids in the rat testis is consistent with the findings of Louie and Dixon (23) in trout, Monesi (24) and Lam and Bruce (25) in mouse, and Loir (26) in the ram. We see no evidence of [^{3}H]-arginine incorporation into somatic histone fractions isolated from highly purified late stage spermatids.

In contrast to the well characterized histones (27), the nonhistone (acid insoluble) proteins are highly heterogeneous and poorly characterized (Figure 5). Between fractions F and G, the heterogeneity appears to drop significantly i.e., several protein bands are absent or greatly reduced. The trend continues in H and I, representing increasingly later stages of spermatid maturation. It is evident from Table I, that the transition from F to G correlates with the morphological change from round spermatids (steps 1-8) to elongated spermatids.

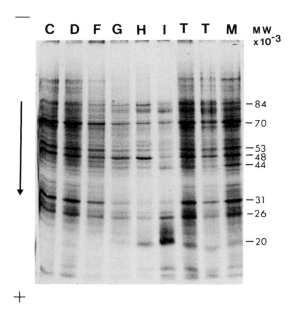

Fig. 5. Nonhistone nuclear proteins from Staput fractions. After extraction of purified nuclei with 0.25 N HCl, the acid insoluble residue was dissolved in 4% SDS, 10% β-mercaptoethanol prior to electrophoresis on an SDS polyacrylamide slab gel (49). The cellular composition of fractions B through I before isolation of nuclei is given in Table I. C, D, F, G, H, and I are Staput fractions from which nonhistone nuclear proteins were isolated; T, trypsin control: whole testis cell suspension prepared with trypsin (15) and held at 4°C for 4 hours to simulate conditions used for Staput fractionation; M, mechanical control: whole testis cell suspension prepared without trypsin. Note that the nonhistones of T and M are virutally identical, indicating no visible degradation of these proteins during preparation of the cell suspension using trypsin. Molecular weight estimates of major proteins are indicated on the right.

Thus it appears that many of the nonhistone proteins are lost in the condensation of chromatin to form a sperm head. This condensation occurs at a stage (steps 1-8) when the full histone complement is still present, and correlates with the repression of RNA synthesis observed in post-meiotic cells (28). Similar changes in the nonhistone proteins have been observed during maturation of avian erythroid cells (29). The presence

of high molecular weight proteins primarily in fractions enriched in pachytene spermatocytes (B and C) is consistent with
the recent observation that high molecular weight nonhistone
proteins do not appear during development of the rat testis
until spermatongonia and spermatocytes have differentiated (30).

Several of the nonhistones in Figure 5 can be tentatively
identified with specific cell types. For example, a protein
band of about 53,000 MW appears to be a component of developing
germ cells up through the early spermatid stage, but is absent
after about step 8. The band at 31,000 MW is associated with
spermatocytes and early spermatids (steps 1-8), while a band
at 48,000 MW is predominantly associated with elongating spermatids (steps 11-15). Bands at 44,000 and 20,000 MW appear to
be major components of late stage spermatids (steps 16-19).
The band at 20,000 MW has the same mobility at TP2 and may,
along with the two proteins below 20,000 MW represent incomplete extraction of basic proteins by 0.25 N HCl. The incorporation of [^{3}H]amino acids into nonhistone protein reveals
significant changes in the amounts and kinds of nuclear proteins being synthesized in different cell types (31). The
highest incorporation occurs in fractions enriched in pachytene
spermatocytes, while little or no incorporation is detectable

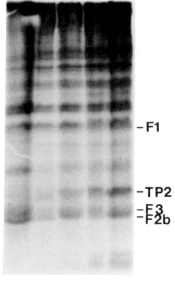

-F1

-TP2

-F3
-F2b

1   2   3   4   5

Fig. 6. Autoradiograph of
^{32}P-labeled nuclear proteins
from rat testis nuclei. Nuclei were isolated as described elsewhere (15) and
dissolved in 4% SDS-10% β-
mercaptoethanol without acid
extraction. Proteins were
separated on a 12% acrylamide slab gel (49). After
staining, the gel was dried
under vacuum and exposed to
X-ray film (Cronex 6, Dupont)
for 11 days.
(1) In vivo. Mechanical
preparation from testis made
60 minutes after intratesticular injection of ^{32}P (800
μCi/testis).

(Continued on next page)

into nonhistone proteins of late spermatids.

We have also examined the phosphorylation of nuclear proteins in rat testis, and find that many are actively phosporylated both in vivo and in vitro (Figure 6). The rapid turnover of nonhistone protein phosphate in the rat testis, however, has prevented us from obtaining reproducible profiles of phosphorylation from individual cell types. Louie and Dixon have demonstrated the phosphorylation of histones and newly synthesized sperm protamine in trout testis. During maturation of the late spermatids the protamine is almost completely dephosphorylated. The low levels of phosphorylation observed by Kadohama and Turkington (30) in mature rat testis may have resulted from the activity of nuclear phosphatases.

Early Development in Sea Urchin.

In view of the evidence pointing to phosphorylation as a significant event in regulation, we wanted to examine this phenomenon in a developmental system where significant changes in transcriptional activity could be expected to occur. Consequently, we have investigated the phosphorylation of nuclear proteins during early development in the sea urchin, Strongylocentrotus purpuratus.

We began by examining the effects of several variables on the kinetics of phosphorylation in whole embryos. The conditions established for in vivo labeling of nonhistone proteins with ^{32}P were 30 minutes incubation at embryo concentrations of 20-40,000 embryos/ml and 2-5 μCi ^{32}P/ml. The available phosphate is taken up from the medium very rapidly and retained by the embryo.

The cumulative incorporation of ^{32}P over the first 3 hours after fertilization was followed as shown in Figure 7A. Although ^{32}P gets into the unfertilized eggs, it does not appear to be incorporated into chromatin proteins. Upon fertilization, however, there is a significant increase in the rate of phos-

---

Fig. 6 continued
(2,3) In vitro. A whole testis cell suspension prepared using trypsin-EDTA (13). After labeling with ^{32}P (50 μCi/ml) for 60 minutes in phosphate-free McCoy's 5A medium (Gibco), the reaction was stopped on ice and nuclei isolated immediately. (4,5) In vitro. Same procedure as 2,3 except that cells were resuspended in phosphate-buffered BSA and held on ice for 2 hours before isolating nuclei.

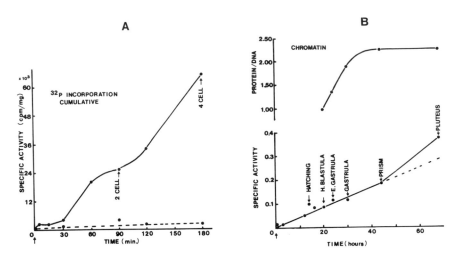

Fig. 7.   Incorporation of ^{32}P into nonhistone protein of
sea urchin chromatin during early development.
    A.   Cumulative incorporation from fertilization to 4-cell
stage.   Eggs were preloaded with ^{32}P for 30 minutes and fertil-
ized at zero time (arrow).   Aliquots were taken at designated
intervals and alkali-labile phosphate determinations (50) were
made on isolated chromatin (51).   The specific activity is
expressed as cpm alkali-labile ^{32}P per mg nonhistone chromatin
protein.   The broken line shows incorporation of ^{32}P into chromatin of unfertilized eggs.
    B.   Incorporation during a 30 minute pulse at various
times from fertilization to pluteus stage.   Embryos were
labeled in sea water with ^{32}P, collected, and chromatin isolated from purified nuclei.   After extraction of basic proteins,
nucleic acids, and lipids, the insoluble residue was assayed
for alkali-labile phosphate.   The specific activity at each
point was divided by the acid soluble pool specific activity to
correct for possible differences in uptake.   The upper portion
of the figure shows the increasing amounts of total protein
associated with DNA in chromatin.

phorylation. Significant increases precede the first cell
division at 90 minutes and the second cell division at 150 minutes. Similar results were obtained by pulse-labeling embryos
for 30 minutes at 30 minute intervals after fertilization.
These data reflect changes in the rate of ^{32}P incorporation
associated with the cell cycle. Consistent with observations
in HeLa cells (32,33), the rate of nonhistone protein phosphorylation is higher during the period of the cell cycle preceding

the actual division of the cell (G2).

The incorporation of ^{32}P was determined at various points from fertilization to the pluteus stage about 72 hours after fertilization (Figure 7B). During this period, the rate of ^{32}P incorporation increases almost linearly. Deviations from linearity are observed during hatching of the blastula stage embryo and again between 45 and 72 hours, during the transition from prism to pluteus. At the time of this later increase, the embryo is just beginning to increase in mass. The upper portion of Figure 7B shows the protein to DNA ratio of chromatin isolated at several points after hatching, and reflects the increasing amounts of nonhistone protein associated with the DNA (34). It seems likely that the rate of incorporation between prism and pluteus remains constant and only appears to increase because the accumulation of nonhistone proteins in the nucleus has leveled off (35). The rate of phosphorylation in prism stage embryos is not significantly lower when RNA synthesis is blocked by actinomycin D or when protein synthesis is inhibited by puromycin (36). In general then, the nonhistone nuclear proteins are rapidly phosphorylated immediately after fertilization and continue to be phosphorylated at an increasing rate throughout embryonic development.

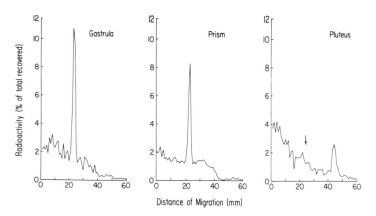

Fig. 8. Profile of radioactivity in sea urchin phosphoproteins after separation on 10% SDS acrylamide gels (38). Embryos at different stages of development were concentrated by centrifugation and labeled for one hour with ^{32}P (39). Radioactivity in each slice is expressed as a percentage of the total radioactivity recovered from the gel. The major phosphorylated peak in gastrula and prism (3,000 cpm and 1,200 cpm respectively) is essentially absent in the pluteus stage (arrow).

We next asked the question: "Are specific nonhistone proteins being phosphorylated at different stages of development?" Phosphoproteins were isolated as described by Gershey and Kleinsmith (37) and separated on SDS-acrylamide gels (38). Using this procedure, we were able to demonstrate the active phosphorylation of specific nonhistone proteins as early as the 32-cell stage (5 hours after fertilization). A typical pattern of labeling emerges by early blastula and is retained through gastula and prism (Figure 8) (39). A single phosphorylated peak dominates the radioactivity profile up through the prism stage.

Repeated efforts to isolate this phosphorylated protein from pluteus stage embryos were unsuccessful until we used filtration instead of centrifugation to collect the plutei before labeling. With the development of a rigid skeleton about 50 hours after fertilization, the embryos become more susceptible to damage during centrifugation. Phosphoprotein preparations from centrifuged plutei lack the highly phosphorylated nuclear protein (39), while plutei concentrated by gentle filtration through 400-mesh nylon exhibit the same phosphorylated protein observed in earlier stages (Figure 9). This protein

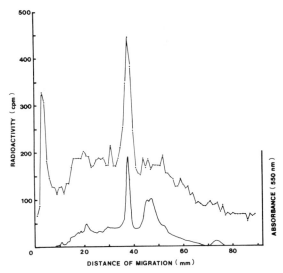

Fig. 9. Profile of radioactivity and absorbance of phosphoprotein isolated from pluteus stage embryos and separated on 10% SDS acrylamide gels. Embryos were concentrated by filtration before labeling for 30 minutes with ^{32}P. Note that the major phosphorylated protein observed in earlier stages is present and phosphorylated in pluteus stage embryos.

TABLE IV

Amino acid	Moles % of total	Approximate residues/molecule
Lysine	5.4	11
Histidine	3.9	8
Arginine	4.2	9
Aspartic acid	7.2	15
Glutamic acid	10.3	22
Threonine	12.0	25
Serine	6.8	14
Proline	10.0	21
Glycine	5.0	10
Alanine	10.0	22
Valine	3.2	7
Methionine	4.8	10
Isoleucine	2.9	6
Leucine	2.9	6
Tyrosine	6.6	14
Phenylalanine	4.4	9

Amino acid composition of the highly phosphorylated protein from prism stage embryos was determined after acrylamide gel electrophoresis. The stained band was cut out and prepared in the presence of β-mercaptoethanol (48). The number of residues/molecule was estimated using a molecular weight estimate of 28,000 based on mobility in SDS gels (38).

has a molecular weight in SDS gels of about 28,000 daltons. The amino acid analysis (Table 4) indicates an unusual composition with threonine, glutamic acid, alanine and proline being the predominant amino acids.

Studies on the kinetics of in vitro phosphorylation were undertaken to further characterize this protein fraction. Since the fraction contains endogenous kinase activity, incubating the total phosphoprotein fraction in the presence of [γ-^{32}P]ATP and 25 mM MgCl$_2$ results in the incorporation of ^{32}P into the phosphoprotein substrate (40). These studies have indicated that the loss of substrate protein in damaged pluteus stage embryos is somewhat selective. The absence of substrate in pluteus, which is apparent from the gel profiles (39), was confirmed by attempting to label the pluteus phosphoprotein in vitro (Figure 10A). The specificity of protein loss became apparent when we found that the phosphoprotein from pluteus,

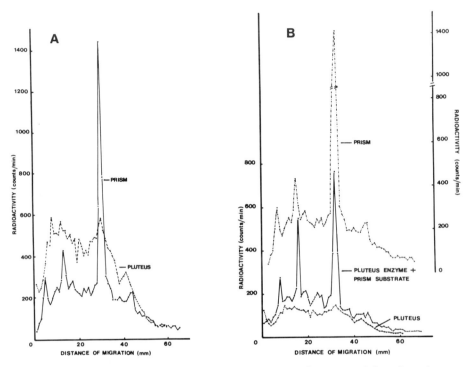

Fig. 10. In vitro phosphorylation of sea urchin phosphoproteins isolated from embryos harvested by centrifugation.
A. The pluteus stage phosphoprotein has no significant
incorporation in the region of the phosphorylated peak from
prism. Protein concentration in the pluteus sample is about 8x
that of the prism, which accounts for the relatively high level
of incorporation by pluteus.
B. Broken lines show incorporation of ^{32}P into prism
(upper) and pluteus (lower) phosphoprotein fractions alone.
Solid line shows the incorporation of ^{32}P into prism substrate
(kinase inactivated by heating) after mixing with the pluteus
phosphoprotein fraction which lacks substrate.

which lacked substrate, was able to phosphorylate prism substrate. That is, the kinase activity was retained in phosphoprotein from damaged plutei even though the substrate had been
destroyed.

Heating the sea urchin phosphoproteins at 60° for 3 minutes destroys more than 80% of the endogeneous kinase activity

TABLE V

	Counts/minute		
	Observed	Expected	Percent
Gastrula	230,000		100
Prism	295,400		100
Pluteus	18,000		100
Heated Prism	54,300		18
Heated Pluteus	3,000		17
Prism + Heated Pluteus	101,200	149,200	68
Pluteus + Heated Prism	47,500	36,200	131

The kinase activity in phosphoprotein preparations from 3 stages of development were assayed at pH 7.5 in a reaction volume of 0.3 ml containing: 13 μmoles Tris-HCl, pH 7.5; 3.5 nmoles of $[\gamma-^{32}P]$ATP (700 μCi/μmole); 7.5 μmoles $MgCl_2$, and 35 μg of phosphoprotein. After incubation for 12 minutes at 37°C, the reaction was terminated by the addition of cold 25% TCA - 3% sodium pyrophosphate. Precipitates were collected on millipore filters presoaked in 1 mM ATP. Each filter was washed 3x with 5% TCA - 1.5% pyrophosphate, 2x with 95% ethanol, air dried, and counted.

Heating at 60°C for 3 minutes destroys more than 80% of the kinase activity. In mixing experiments, the kinase activity of the pluteus stage phosphoprotein was observed to phosphorylate substrate from prism stage phosphoprotein. Expected values were calculated as one-half the sum of the counts incorporated by fractions individually.

(Table V). The combination of pluteus phosphoprotein with heated prism phosphoprotein incorporates 30% more ^{32}P than expected from the sum of the counts incorporated by each fraction independently. Thus, the prism kinase cannot significantly phosphorylate the pluteus phosphoprotein because the substrate has been destroyed, but the pluteus phosphoprotein has kinase activity which is capable of phosphorylating the prism substrate. The identity of the substrate being phosphorylated by the kinase from pluteus stage embryos was confirmed by SDS acrylamide gel electrophoresis (Figure 10B).

The selective loss of the nonhistone chromatin phospho-protein indicated by these data could occur in a variety of ways. The most likely explanation is that the substrate protein is especially sensitive to nonspecific proteolytic activity released by the damaged cells. Alternatively, if the substrate protein is rapidly turned over in the intact embryo, dissociation of the cells could stop its synthesis, leaving the normal degradation pathway uncontrolled.

We are currently engaged in the fractionation of phospho-protein kinases isolated from embryos at different developmental stages. Using the procedure described by Kish and Kleinsmith (41) for beef liver, protein kinase activity from sea urchin embryos may be separated into 10-12 different fractions. Preliminary results indicate that the profile of enzyme activity from blastula stage embryos is markedly different from late gastrula and pluteus.

During early development the nonhistone proteins are rapidly phosphorylated. Significant increases in the rate of phosphorylation appear to be associated with fertilization and hatching. The functional significance of specific nuclear protein phosphorylation in developmental systems remains to be determined. Phosphorylation should continue to be a useful tool in probing the role of nuclear proteins in gene regulation.

CONCLUSION

Our present understanding of the genetic regulatory mechanisms operating in eukaryotic cells is extremely limited. Selective fractionation of the nonhistone proteins together with further studies on their synthesis and modification should help in identifying and characterizing specific chromatin protein components. The development of hybridization techniques using cDNA to detect specific mRNA molecules makes it possible to assay the effect of specific proteins on transcription. Application of these techniques to the analysis of differentiating cells will provide considerable insight into the molecular mechanisms underlying differentiation.

Acknowledgements

We are grateful to Dr. B. D. Burleigh for doing the amino acid analyses for us. Supported by USPHS Grants HD-05803, CA-06294, a Contract FDA-73-204 from NCTR, Robert A. Welch Foundation Grant G138 and a grant from the Population Council, New York.

References

1. F. Jacob. Science 152:1470 (1966).

2. R. J. Britten and E. H. Davidson. Science 165:349 (1969).

3. A. Tyler. In: "Analysis of Development", B. H. Willier, P. A. Weiss and V. Hamburger, eds. W. B. Saunders, Philadelphia, p. 170 (1955).

4. M. F. Lyon. Nature 190:372 (1961).

5. O. A. Schjeide and J. deVellis. In: "Cell Differentiation", O. A. Schjeide and J. deVellis, eds. Van Nostrand Reinhold, New York, p. 7 (1970).

6. P. R. Harrison, A. Hell, G. D. Birnie and J. Paul. Nature 239:219 (1972).

7. R. S. Gilmour and J. Paul. Proc. Nat. Acad. Sci. USA 70: 3440 (1973).

8. G. S. Stein, T. C. Spelsberg and L. J. Kleinsmith. Science 183:817 (1974).

9. G. Giudice. "Developmental Biology of the Sea Urchin Embryo," Academic Press, New York (1973).

10. K. Marushige and G. H. Dixon. Develop. Biol. 19:397 (1969).

11. M. Courot, M. T. Hochereau-de Reviers and R. Ortavant. In: "The Testis", Vol. 1 p. 339, A. D. Johnson, W. R. Gomes and N. L. Vandemark, eds. Academic Press, New York.

12. D. M. K. Lam, R. Furrer and W. R. Bruce. Proc. Nat. Acad. Sci. USA 65:192 (1970).

13. M. L. Meistrich and V. W. S. Eng. Exp. Cell Res. 70:237 (1972).

14. S. R. Grimes, R. D. Platz, M. L. Meistrich and L. S. Hnilica, submitted for publication.

15. R. D. Platz, S. R. Grimes, M. L. Meistrich and L. S. Hnilica. J. Biol. Chem., in press (1975).

16. S. R. Grimes. Ph.D. Thesis, Department of Biochemistry, University of North Carolina (1973).

17. W. S. Kistler, M. E. Geroch and H. G. Williams-Ashman. J. Biol. Chem. 248:4532 (1973).

18. R. E. Branson, S. R. Grimes, G. Yonuschot and J. L. Irvin. Arch. Biochem. Biophys., in press (1975).

19. D. E. Comings and T. A. Okada. Adv. in Cell and Mol. Biol. 2:309 (1972).

20. J. R. Davis and G. A. Langford. In: "The Testis", Vol. 2 p. 259, A. D. Johnson, W. R. Gomes and N. L. Vandemark, eds. Academic Press, New York (1970).

21. Y. Marushige and K. Marushige. J. Biol. Chem. 250:39 (1975).

22. J. P. Coelingh, C. H. Monfoort, T. H. Rozijn, J. A. G. Leuven, R. Schiphof, E. P. Steyn-Parve, G. Braunitzer, B. Schrank and A. Ruhfus. Biochim. Biophys. Acta 285:1 (1972).

23. A. J. Louie and G. H. Dixon. J. Biol. Chem. 247:5490 (1972).

24. V. Monesi. Exp. Cell Res. 39:197 (1965).

25. D. M. K. Lam and W. R. Bruce. J. Cell. Physiol. 78:13 (1971).

26. M. Loir. Ann. Biol. Anim. Biochim. Biophys. 12:411 (1972).

27. L. S. Hnilica. "The Structure and Biological Functions of Histones," Chemical Rubber Company Press, Cleveland, Ohio (1972).

28. V. Monesi. J. Reprod. Fert., Suppl. 13:1 (1971).

29. A. Ruiz-Carrillo, L. J. Wangh, V. C. Littau and V. G. Allfrey. J. Biol. Chem. 249:7358 (1974).

30. N. Kadohama and R. W. Turkington. J. Biol. Chem. 249:6225 (1974).

31. R. D. Platz, S. R. Grimes, M. L. Meistrich and L. S. Hnilica, manuscript in preparation.

32. R. D. Platz, G. S. Stein and L. J. Kleinsmith. Biochem. Biophys. Res. Comm. 51:735 (1973).

33. J. Karn, E. M. Johnson, G. Vidali and V. G. Allfrey. J. Biol. Chem. 249:667 (1974).

34. K. Marushige and H. Ozaki. Develop. Biol. 16:474 (1967).

35. R. L. Seale and A. I. Aronson. J. Mol. Biol. 75:633 (1973).

36. R. D. Platz and L. S. Hnilica, manuscript in preparation.

37. E. L. Gershey and L. J. Kleinsmith. Biochim. Biophys. Acta 194:331 (1969).

38. K. Weber and M. Osborn. J. Biol. Chem. 244:4406 (1969).

39. R. D. Platz and L. S. Hnilica. Biochem. Biophys. Res. Comm. 54:222 (1973).

40. L. J. Kleinsmith and V. G. Allfrey. Biochim. Biophys. Acta 175:123 (1969).

41. V. M. Kish and L. J. Kleinsmith. J. Biol. Chem. 249:750 (1974).

42. R. Miller and R. Phillips. J. Cell. Physiol. 73:191 (1969).

43. M. L. Meistrich, W. R. Bruce and Y. Clermont. Exp. Cell Res. 79:213 (1973).

44. J. C. Vaughn. J. Cell Biol. 31:257 (1966).

45. S. Panyim and R. Chalkley. Arch. Biochem. Biophys. 130:337 (1969).

46. J. P. Coelingh, T. H. Rozijn and C. H. Monfoort. Biochim. Biophys. Acta 188:353 (1969).

47. C. H. W. Hirs. In: "Methods in Enzymology", Vol. XI, C. H. W. Hirs, ed. Academic Press, New York, p. 197 (1967).

48. L. L. Houston. Anal. Biochem. 44:81 (1971).

49. U. K. Laemmli. Nature 227:680 (1970).

50. L. J. Kleinsmith, V. G. Allfrey and A. E. Mirsky. Proc. Nat. Acad. Sci. USA 55:1182 (1966).

51. T. C. Spelsberg and L. S. Hnilica. Biochim. Biophys. Acta 228:202 (1971).

THE DEPOSITION OF HISTONES ONTO REPLICATING CHROMOSOMES

Vaughn Jackson, Daryl K. Granner and Roger Chalkley
Department of Biochemistry
The University of Iowa
Iowa City, Iowa 52242

Abstract

The deposition of histones onto the replicating mammalian chromosome has been studied. We have utilized a technology which avoids intermolecular randomization of histones and can successfully account for the concommitant distribution of non-histone proteins. A series of experimental protocols permit a distinction between the three major modes of distribution; namely, conservative, in which all newly synthesized histones become associated with one daughter chromosome; semi-conservative in which histones maintain an interaction with one or other of the parental DNA strands; or random deposition. We will argue that when precautions are taken to avoid artifacts, particularly re-organization, then the distribution of both lysine-rich and arginine-rich histones follows a random mode.

The deposition of newly synthesized histones in the replicating nucleus is a complex problem. An approach to this area has consisted of asking how pre-existing histones distribute themselves with respect to specific strands of DNA. In general, three models are possible: (A) conservative, in which all the old histone is associated with one daughter DNA molecule and all the new histone with the other DNA molecule, (B) semi-conservative, in which old histones retain an interaction with one specific strand in each daughter DNA molecule and new histones interact specifically with the other strand, and (C) a fully random mode. Autoradiographic studies by Prescott and Bender (1) following a pulse of labeled amino acids in dividing amoebae indicated that in subsequent generations the radiolabel is uniformly associated with all chromatids. Unfortunately, this approach cannot distinguish between histones and nonhistone proteins. However, if one assumed that histones were selectively transferred into specific daughter cells, it would be necessary to invoke the argument that a corresponding amount of radioactivity in nonhistone is transferred into the other daughter cells. The results of Prescott and Bender are

93

entirely consistent with a random mode of histone deposition, but of course do not prove this point.

More recently, Tsanev and Russev (2) have argued that at replication old histones remain associated with the old DNA strands and that newly synthesized histone is associated with the new DNA strand. This analysis is critically dependent upon the ability to separate small nucleoprotein molecules from dissociated histone and nonhistone proteins, and because of the large amount of background radioactivity in this region of the sucrose gradients, it is difficult to conclude unambiguously that histones do indeed follow a semiconservative mode of deposition.

We felt, therefore, that it would be appropriate to adopt a completely different approach utilizing incorporation of density label into DNA (3,4) and analyzing fixed (5,6) chromatin preparations in CsCl density gradients. Our results indicate that histones are distributed on the replicating DNA in a random manner.

MATERIALS AND METHODS

Labeling of HTC Cells with [^{3}H]lysine, [^{3}H]tryptophan and [^{3}H]arginine

400 ml of HTC (Hepatoma Tissue Culture) cells in mid-log phase (4 x 10^5 cells/ml) were labeled with the appropriate radioactive amino acid (200 μc, New England Nuclear) in the presence (Expt. 1) or absence (Expt. 3) of 5'-iododeoxyuridine* (Nutritional Biochemicals) at 1 x 10^{-4} M. After 16 hours (one cell generation), 200 ml of the cells were harvested and the remaining 200 ml were centrifuged (200 x g) and resuspended in 400 ml of fresh medium (Swims S-78) containing 1 x 10^{-5} M thymidine (Expt. 1) or 1 x 10^{-4} M IdU (Expt. 3). After continued growth for an additional 16 hours, 200 ml of cells were harvested and the remaining 200 ml replaced by 400 ml of fresh medium containing 1 x 10^{-4} M thymidine. This process was repeated until all of the chases of the labeled amino acid had been completed. In Expt. 2, HTC cells (400 ml) were preincubated with 1 x 10^{-4} M IdU for 16 hours prior to labeling with the amino acid.

When the cells were harvested, either they were directly frozen (when chromatin was to be isolated by our standard procedure) or, when treated following the Hancock procedure (8),

---

*abbreviation used: 5'-iododeoxyuridine = IdU

they were washed once in an equal volume of 0.1 M sucrose, 0.2 mM Na_2HPO_4, pH 7.4, at 4°C and then frozen in dry ice-ethanol.

Preparation of Nuclei, Isolation of Chromatin and Analysis of CsCl Gradients

Preparation of chromatin by our standard procedure was as previously described (4) and involves isolation of nuclei in the presence of 1% Triton X-100, 0.25 M sucrose, 0.01 M $MgCl_2$, 0.01 M Tris HCl, 0.05 M $NaHSO_3$, pH 6.5. Preparation of chromatin by the procedure of Hancock (8) involves an initial homogenization in 0.1 M sucrose, 0.2 mM Na_2HPO_4, pH 7.4, followed by several washes in 0.5% Nonidet P-40, 0.2 mM EDTA, pH 8.0. The chromatin pellet was then washed several times in 0.2 mM EDTA, pH 8.0, to remove the detergent and finally once with distilled water to form the chromatin gel. The gel was then vigorously sheared for 1 min at 4°C in a Virtis Model "45" homogenizer at maximum shear force, followed by centrifugation at 27,000 x g for 20 min to remove membrane fragments. This supernatant, which contains the solubilized chromatin, was dialyzed against 1×10^{-5} M triethanolamine, pH 7.4 (TEA), fixed in 1% formaldehyde for 2 hrs and redialyzed against 1×10^{-5} M TEA to remove excess formaldehyde. The chromatin was mixed with guanidine hydrochloride (2.1 g) (Heico, Inc.) and CsCl (1.5 g) and adjusted to a final volume of 5.5 ml with 0.1 M Tris HCl, pH 8.0. After centrifugation at 35,000 rpm in a Beckman SW 50.1 rotor for 72 hrs at 4°C, 20-drop fractions were collected and counted in Bray's solution (9) using a Unilux III scintillation counter. Equivalent counts of [^{3}H]amino acids were applied to each gradient, and the [^{3}H]amino acid/[^{14}C]thymidine ratio was kept constant to ensure equivalent double label analysis in all experiments. The distribution of radioactivity is independent of DNA concentrations between 25-500 μg in CsCl-GuCl gradients.

In order to analyze the amount of DNA containing IdU, the solubilized chromatin (prior to fixation) was dissolved with 6.80 g CsCl into a final 7.0 ml volume of 0.1 M Tris HCl, pH 8.0, and centrifuged at 40,000 rpm in a Beckman #50 rotor for 48 hrs at 4°C. Twenty five-drop fractions were collected and the absorbance at 260 nm determined.

The percentage of [^{3}H]lysine present as nonhistone protein was determined after histones were extracted from the solubilized chromatin and analyzed on 15% acrylamide-2.5 M urea gels (7). These gels were sliced and counted as previously described (7).

TABLE I

DISTRIBUTION SCHEMES FOR HISTONE DEPOSITION

Model	Experiment 1	Experiment 2	Experiment 3
A. Conservative			
1. old histone on older DNA strand	D, N, N	D, D, D	N, D, N, N
2. old histone on newer DNA strand	D, D, N	N, N, N	N, D, D, N
B. Semi-conservative			
3. old histone on old DNA strand	D, D, D,	50% D, N, N	N, D, N, N
4. old histone on new DNA strand	D, N, N	50% D, 50% D, N	N, D, D, N
5. old histone on old DNA strand (but strand switch)	D, N, N	50% D, 50% D, N	N, D, D, N
6. old histone on new DNA strand (strand switch)	D, D, D	50% D, N, N	N, D, N, N
7. old histone on new/old strand alternating	D, D, N	50% D, 25% D, N	D, N, N, N
C. Random			
8. random with respect to DNA strands and molecules	D, 50% D, 25% D	50% D, 25% D, 12 1/2% D	N, D, 50% D, 25% D

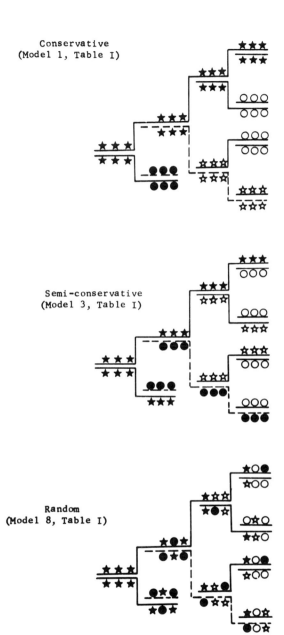

Fig. 1.   Caption on next page.

RESULTS

We can conceive of three ways whereby newly synthesized histone could deposit on DNA: conservative, semiconservative and random deposition. The conservative and semiconservative modes of deposition can be subdivided into several different forms (Table I).

In Figure 1 we see an example of conservative deposition in which the newly synthesized histone is deposited on the daughter DNA molecule containing the younger parental DNA strand; conversely, the pre-existing histone is distributed so that it is in association with the daughter DNA molecule containing the older strand from the parent DNA (Table I, Model 1). Although it is not shown in Figure 1, there is another mode of conservative deposition in which the newly synthesized histone becomes associated with a DNA molecule containing the older parental strand (Table I, Model 2). One possible form of the semiconservative mode is shown in Figure 1. In this example the pre-existing histone at the time of replication distributes itself equally onto both daughter DNA molecules, and upon subsequent division is always associated with the same DNA strand (Table I, Model 3). This is the model for histone deposition proposed by Tsanev and Russev (2). We can conceive of five variants of this mode of deposition, depending upon whether incoming histone is associated with a pre-existing or a newly synthesized DNA strand and whether the histones can switch strands at replication. The third model for organization of histones during chromosome replication is the random approach (Table I, Model 8), which should be adequately defined in Figure 1.

We have conducted a series of three types of experiments designed to distinguish among these various possibilities. The strategy involved is discussed below, with specific reference to the results one might predict if the semiconservative model presented in Table II were to be that most accurately depicting events occuring in the cell.

Fig. 1. Typical models for organization of histones during chromosome replication. First generation of histones, ✷✷; 2nd generation, ●●; 3rd generation, ✩✩; 4th generation, 0 0. We have assumed there is no significant turnover of histones during this time period in HTC cells (7). Each branch point represents the completion of a full cell generation. The dotted line represents the subsequent distribution of a density label incorporated during the 2nd generation.

TABLE II

PREDICTED DENSITY OF [^{3}H]LYSINE-LABELED HISTONE FOLLOWING SEMICONSERVATIVE DEPOSITION MODE[d]

EXPERIMENT	PULSE	1st CHASE	2nd CHASE	3rd CHASE
1. IdU[a] + [^{3}H]lysine	D[b]	D	D	-
2. IdU one generation then [^{3}H]lysine	(50% D, 50% N)	N[c]	N	-
3. [^{3}H]lysine one generation then IdU one generation	N	D	N	N

[a] iododeoxyuridine

[b] D = high density nucleohistone containing one strand labeled with iododeoxyuridine and one strand of normal density DNA.

[c] N = normal density nucleohistone

[d] Mode B-3 (Table 1). Semiconservative deposition of Figure 1.

Experiment 1 involves growing HTC cells in the presence of both [^{3}H]lysine and iododeoxyuridine (IdU) for 16 hrs (one full cell cycle). Under these conditions all of the newly synthesized DNA strands in the chromatin contain the density label, and the newly synthesized histone contains [^{3}H]lysine. This is the pulse period shown for Experiment 1 in Table II. Both the IdU and [^{3}H]lysine are then replaced by fresh medium containing (1 x 10^{-5} M) thymidine and chased for 16 hrs and 32 hrs, respectively. Under these conditions the chase period continues for two full rounds of replication. If the semiconservative mode of deposition (shown in Table II) were to be correct, we predict that after the pulse of [^{3}H]lysine and IdU, the newly synthesized [^{3}H]lysine-labeled histone will be deposited on DNA molecules containing the heavy base and the nucleoprotein will therefore be more dense (D) than normal nucleohistone (N). After the first chase (16 hrs) the [^{3}H]lysine-containing histone is still associated with DNA containing the heavy base. Likewise after the second chase of 16 hrs the [^{3}H]histone would still be associated only with those chromosomes containing the density labeled DNA.

The second set of experiments involves incubating cells with IdU for 16 hrs and then after removal of the IdU, treating the cells with [³H]lysine for 16 hrs. This latter incubation is the pulse of Experiment 2 in Table II. Under these conditions the newly synthesized histone is now deposited on DNA, half the molecules of which contain one dense strand and the other half of the DNA molecules contain both strands of normal density. Thus at the conclusion of the [³H]lysine pulse period the model of Table II predicts that radiolabel will be associated with two classes of nucleoprotein, one dense and one of normal density; this is designated 50% D, 50% N in Table II. Following a chase period (no [³H]lysine or IdU in the medium) for 16 hrs, the model of Table II predicts that the [³H]lysine will no longer be associated with dense DNA and the radiolabeled nucleoprotein will have the same density as normal nucleohistone (N). Similarly during a second chase of 16 hrs this model predicts that the [³H]lysine-labeled nucleohistone will continue to be of normal density (N, Table II).

The third set of experiments involves incubating cells with [³H]lysine for 16 hrs (pulse period of Experiment 3, Table II), and after removal of [³H]lysine, treating the cells with IdU for 16 hrs (first chase of Experiment 3, Table II). The [³H]lysine after the pulse should be of normal density (N), and after the first chase it is expected to be dense (D), as it is now associated with DNA molecules all of which contain the dense base. However, during the second chase of 16 hrs initiated immediately after removal of IdU, the [³H]histone will be on the DNA strand which does not contain the density label, and therefore the labeled nucleoprotein will be of normal density (N). One would predict that during a third chase of 16 hrs the [³H]lysine will continue to be associated with chromatin which is of normal density (N). Applying similar reasoning to the 8 models of Table I leads to a series of specific predictions concerning the association of [³H]labeled histones with density-labeled DNA. The predictions documented in Table I are sufficiently unambiguous that one can discriminate clearly between the three general models for deposition and, to a degree, between the variants within each general model.

In order to measure the density changes due to the presence of IdU, the nucleohistone isolated from these cells was fixed with formaldehyde and then examined on CsCl density gradients. Figure 2 shows the density distribution of nucleohistone containing both IdU and [³H]thymidine compared to nucleohistone of normal density containing [¹⁴C]thymidine. The density difference is sufficiently large that one can readily distinguish between normal (N) and dense (D) nucleohistone.

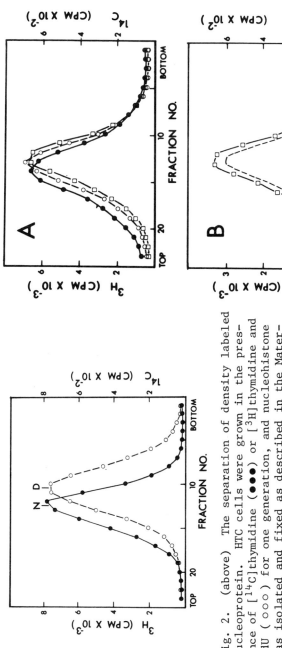

Fig. 2. (above) The separation of density labeled nucleoprotein. HTC cells were grown in the presence of [^{14}C]thymidine (●●●) or [^{3}H]thymidine and IdU (○○○) for one generation, and nucleohistone was isolated and fixed as described in the Materials and Methods. The fixed nucleoproteins were mixed and analyzed on a CsCl density gradient. After equilibrium was reached, fractions were collected and counted.

Fig. 3. (at right) Determination of distribution of [^{3}H]lysine in histone in the presence of nonhistone protein. A. HTC cells were grown in the presence of [^{3}H]tryptophan, [^{3}H]lysine or [^{14}C]thymidine for 16 hrs. Chromatin was isolated and fixed from each source. Chromatin containing [^{14}C]thymidine (-□□-) was mixed with either [^{3}H]-lysine chromatin (○○○) or the [^{3}H]tryptophan (●●●) chromatin and was centrifuged in CsCl as described in the Methods. (See next page for caption 3B)

The experiments to analyze histone distribution during replication require that we have a knowledge of the distribution of radiolabeled histone in the CsCl gradients of fixed chromatin. Unfortuantely nonhistone proteins not only incorporate [^{3}H]lysine, but also are fixed within the chromosomal complex with some efficiency. It is therefore necessary to apply a correction for nonhistone protein distribution in order to assay precisely for the position of [^{3}H]histone within the density gradients. This correction can be performed by exploiting the observation that histones do not contain tryptophan, which allows one to use this amino acid as a marker for nonhistone protein density distribution in CsCl density gradients. This approach is illustrated by the results presented in Figure 3A, which show the distribution of [^{3}H]tryptophan in nonhistone proteins and of [^{3}H]lysine in both nonhistone proteins and in histones. The amount of [^{3}H]lysine which is present in nonhistone proteins is determined directly on all samples before fixation and centrifugation using acid solubility and gel analysis. In general, we have observed that about 50% of the label is nonhistone protein and some 50% is found in histones. A typical determination is shown in the legend of Figure 3. We then compare the density distribution profile of [^{3}H]tryptophan (nonhistone) to that of the [^{3}H]lysine (nonhistone + histone), using a common internal standard ([^{14}C]-thymidine in nucleohistone added to each sample). The area under the curve showing the distribution of [^{3}H]tryptophan was adjusted to be a required fraction (i.e. the fraction of the total [^{3}H]lysine counts due to nonhistone protein) of the area of the [^{3}H]lysine density distribution using a Dupont curve analyzer so that the shape of the curve is maintained faithfully even though the area is changed. The adjusted nonhistone [^{3}H]tryptophan curve is then subtracted from the (nonhistone + histone) [^{3}H]lysine curve to give the histone distribution curve in the CsCl density gradient. The data shown in Figure 3A were corrected in this way and are shown in Figure 3B.

In order for these experiments to be meaningful, it is important to demonstrate that histones are not redistributed between DNA molecules during the preparative procedure. To

---

Fig. 3B. Corrected distribution of [^{3}H]lysine in histone (---) relative to [^{14}C]thymidine-containing chromatin (-□ □-). The correction is described in the text and critically dependent upon a determination of the amount of [^{3}H]lysine in both histone and nonhistone protein. We measure the amount of [^{3}H]lysine in acid-insoluble protein and that fraction of acid-soluble radioactivity which is not specifically in histone molecules as determined by gel electrophoresis.

test for this possibility, intact cells previously labeled with [^{3}H]lysine for one cell cycle were mixed with a five-fold excess of cells grown for two generations in the presence of IdU. The mixture of two cell populations was homogenized and chromatin isolated. The nucleohistone was isolated either by our standard procedure or by a procedure recently devised by Hancock (8) to avoid randomization of histone. The chromatin was fixed in 1% formaldehyde and analyzed on CsCl density gradients. Controls for normal density and for dense chromatin were obtained by growing cells for two generations in the presence of either [^{3}H]lysine alone (N) or [^{3}H]lysine plus IdU (D). These control samples were centrifuged in the same rotor, but in different tubes. Alignment of the experimental and control density distribution was achieved by the addition of fixed chromatin of normal density containing [^{14}C]thymidine to all samples. A parallel experiment utilizing [^{3}H]tryptophan was also incorporated into this analysis with the results shown in Figure 4A-F. Figure 4A and B shows the observed redistribution of [^{3}H]lysine and [^{3}H]tryptophan when isolated by our normal isolation procedure. Figure 4D and E shows the observed redistribution when isolated by the procedure suggested by Hancock. When the distribution in nonhistone protein ([^{3}H]tryptophan) is subtracted from the [^{3}H]lysine distribution as described in the text for Figure 3, then the observed histone redistribution during our standard procedure (Figure 4C) and Hancock's procedure (Figure 4F) can be determined.

It is apparent that a small degree of histone redistribution takes place during our normal isolation procedures, but that if divalent cations are excluded from the isolation media as suggested by Hancock (8), essentially no intermolecular reorganization of histone occurs. Reorganization with respect to strand of a given histone may conceivably occur, but this will not affect the analysis as outlined in Figure 1 and Table I. The rest of the experiments to be described utilize the Hancock isolation procedure.

For this experimental technique to be successful it is important that the pulse and chase parts of the experiments do indeed encompass, rather precisely, one cell generation, and that DNA synthesis continues unaffected by the various experimental manipulations. In all the experiments to be described we have checked these parameters by analyzing the efficiency and time dependence of incorporation of density label into DNA. Samples of unfixed chromatin were mixed with CsCl solutions so that DNA of different densities could be separated. A typical example is shown in Figure 5 for Experiment 1 of Table II. We see that after one generation of incorporation of IdU, essen-

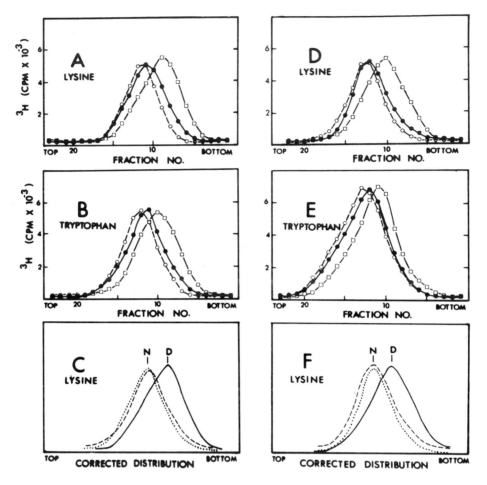

Fig. 4A-F (above).   Captions on next page.

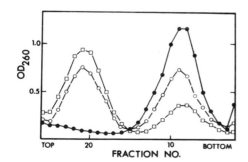

Fig. 5.   Caption on next page.

Fig. 4.   Effect of mode of isolation on randomization of histones.   HTC cell chromatin of normal density containing [^{3}H]-amino acid (-●●●-) was co-isolated with a five-fold excess of chromatin containing IdU.   A control chromatin of normal density (-ooo-) and chromatin of high density (-□□□-), both containing [^{3}H]amino acids, were also isolated separately.   The three experiments were analyzed in the same rotor head on CsCl density gradients and positioned relative to one another by a control, fixed [^{14}C]thymidine chromatin standard.   The distribution of [^{3}H]lysine in histone was obtained by a correction of the type shown in the text and in the legend to Figure 3. (A-C), Isolation of mixed cells in the presence of Mg^{++}; (D-F), Isolation of mixed cells using Hancock's procedure (8).   In the panels showing the corrected histone distributions (C and F) the following designations are used: (....) histone distribution in normal density chromatin; (——), histone distribution in dense (IdU) chromatin and (----) histone distribution in the mixing experiment in which normal density, [^{3}H]-labeled material was mixed with a five-fold excess of IdU-containing cells and chromatin isolated from the mixed cells following the two procedures, C and F.

Fig. 5.   Precision of measurement of cell generation time. HTC cells were pulsed with IdU for 16 hrs (as a part of the protocol for Experiment 1), and an aliquot of cells collected. The remainder of the cells were washed and permitted to grow in fresh medium for either 16 or 32 hrs.   Chromatin was isolated and applied to CsCl solutions without fixation so that the distribution of DNA could be measured.   DNA in the cells after 16 hrs density pulse (-●●●-), after 16 hrs of chase in the absence of density label (-ooo-), after 32 hrs of chase (-□□□-).

Fig. 6.   Distribution of [^{3}H]lysine and [^{3}H]tryptophan relative to density labeled DNA during Experiment 1.
    A.   The distribution of [^{3}H]lysine in both histones and nonhistone protein following a pulse of (IdU + [^{3}H]lysine), (-●●●-); after one generation chase, (-□□□-); after two generations chase, (-ooo-).
    B.   The distribution of [^{3}H]tryptophan in nonhistone protein during the IdU + [^{3}H]tryptophan pulse, (-●●●-); after one generation chase, (-□□□-); after two generations chase, (-ooo-).   All samples were aligned with respect to a fixed chromatin sample containing [^{14}C]thymidine (not shown).

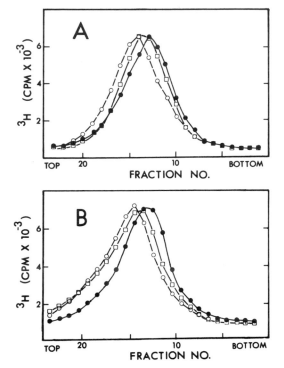

Fig. 6. Caption on previous page.

tially all the DNA is in the denser peak; after one generation of subsequent growth in fresh medium, we see that 50% of the DNA is of normal density and 50% is of the higher density. A second generation produced 75% of the DNA of normal density and 25% of high density. This is precisely the distribution expected for DNA which was labeled with IdU for one generation and chased for either one or two generations. Evidently these cells grow well in the presence of 1×10^{-4} M IdU and there is no significant effect on the cell generation time.

Having established the basic technology, we were in a position to perform the three types of experiments discussed in Table II and analyzed in detail in Table I. The raw data for the first experiment, uncorrected for nonhistone protein distribution, are shown in Figure 6; application of the correction leads to the data of Figure 7. In the first experiment both [^{3}H]lysine and IdU were present for one generation. As shown in Figures 6 and 7A, after this pulse period both the [^{3}H]lysine and nonhistone proteins are distributed fully on dense DNA as expected, and as such this experiment serves as a control for the position of dense nucleoprotein within the CsCl

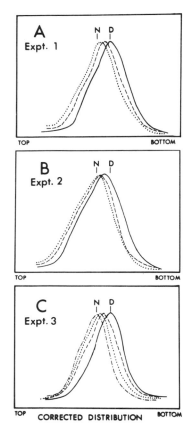

Fig. 7. Distribution of [³H]-histone with respect to density-labeled DNA.
A. Corrected histone distribution during Experiment 1 (see text). After 16 hr pulse of IdU + [³H]lysine, (———); after 16 hr chase, (----); after 32 hr chase, (....).
B. Corrected histone distribution during Experiment 2 (see text). After 16 hr pulse in IdU, followed by 16 hr chase in [³H]-lysine, (———); after 16 hr chase, (----); after second 16 hr chase, (....).
C. Corrected histone distribution during Experiment 3 (see text). After 16 hr [³H]lysine pulse, (.-.-.); after 16 hr chase in IdU, (———); after 16 hr chase no label, (----); after final 16 hr chase, (....). All samples were aligned with [¹⁴C]thymidine in control fixed chromatin. 'N' refers to the peak of [³H]histone distribution on chromatin containing no density label: 'D' refers to the peak of [³H]histone distribution on chromatin, the DNA molecules of which all contain one strand which is fully density labeled with IdU.

density gradient. After one full cell cycle in the absence of either label the [³H]lysine is distributed uniformly, midway between normal density and the dense nucleoprotein. This is the expected distribution for a mixture of equal numbers of molecules of dense and normal density material, both containing equal amounts of [³H]lysine. After a second generation the [³H]lysine distribution is shifted towards lower density but, none the less, it is still situated slightly to the dense side of a normal density nucleoprotein. This is expected for a population of three parts normal density nucleoprotein and one part dense. If the mode of distribution proposed by Tsanev and Russev (2) were correct, the histone should have remained associated with dense DNA throughout all phases of the experiment (see Table II). In the second experiment (Figure 7B) the density label was added for one generation and then replaced by [^{3}H]lysine for one cell generation. Immediately after the [^{3}H]lysine pulse, 50% of all the DNA molecules are density labeled. [^{3}H]histones are distributed half-way between the positions of normal density and dense chromatin. After a chase period the histone distribution shifts to lighter regions; however, even after a second chase period normal density distribution of [^{3}H]histone is not fully attained, and we therefore conclude that we are seeing a continual dilution of radiolabeled histones associated with the density labeled DNA molecules.

The third experiment (Figure 7C) consisted of an initial pulse of [^{3}H]lysine, after which the histone is obviously on normal density nucleoprotein and as such this forms a control for the distribution of normal density nucleoprotein. We next initiated a chase in which the radiolabel was replaced by IdU (at this point the histone is completely shifted onto dense DNA). Additional chase periods ensued, both of one cell generation in time span. After the first such chase in the total absence of label, the radioactive histone is found distributed half-way between normal and dense DNA molecules, while after the second chase period 75% of the [^{3}H]histone is on DNA molecules of normal density.

The entire approach was repeated with [^{3}H]arginine, and strictly analogous data were obtained. The results we have obtained are summarized in Table III. We conclude that the arginine-rich histones deposit in the same manner as the lysine-rich histones. These results are consistent with a random mode of deposition as outlined in Table I.

TABLE III

SUMMARY OF RESULTS CONCERNING HISTONE DISTRIBUTION

	EXPERIMENT	PULSE	1st CHASE	2nd CHASE	3rd CHASE
1.	IdU + [^{3}H]lysine	D	50% D*	25% D	-
2.	IdU one generation then [^{3}H]lysine	50% D	25% D	12 1/2% D	-
3.	[^{3}H]lysine one generation then IdU one generation	N	D	50% D	25% D

Caption on next page.

108

DISCUSSION

Studies on the distribution of histones during chromosome replication have proven to be technically demanding. It is necessary to avoid histone reorganization. Experiments show that using the Hancock isolation procedure there is no randomization of histones from one nucleoprotein molecule to another either during the isolation of chromatin or during the subsequent shearing. We also note that in HTC cells the nonhistone proteins do undergo a considerable degree of randomization during the isolation procedures. It is also essential to have reasonable estimates of the distribution of nonhistone chromosomal proteins within the chromatin density profile on CsCl gradients, and to have an accurate knowledge of the number of cell generations involved in the period under study. Further, if one wishes to exploit CsCl density gradients, it is necessary to cross-link the histones to DNA without introducing intermolecular interactions between separate nucleoprotein molecules, and to expose the fixed nucleoprotein to the high salt environment without obtaining a concomitant aggregation (guanidine hydrochloride must be present in the gradient) (4).

Utilizing the experimental protocol we have devised, we conclude that histones become associated with newly synthesized DNA in a random manner. Further, as histones do not turn over significantly, we conclude that pre-existing histones are likewise randomly disposed onto the daughter DNA molecules during replication. The results also preclude a model whereby an incoming histone could interact with either DNA strand at random but that once this interaction is established the DNA and histone could not separate. This conclusion is in contradiction to that obtained by Tsanev and Russev (2). However, these authors utilized a heterogeneous population of dividing and non-dividing cells (regenerating liver) of uncertain frequency and time-length of generation. Their isolation procedures may well have been conducive to reorganization, and rigorous attempts were not made to account for the distribution of nonhistone proteins. On the other hand Prescott and Bender (1) labeled the entire chromosomal proteins and studied the distri-

Table III. * The notation 50% D implies that 50% of the [3H]-lysine is associated with the denser chromatin (one strand labeled with IdU and one strand normal density) and that 50% of the radiolabel is associated with chromatin entirely lacking IdU. Obviously the other possible meaning, namely that all the [3H]histone is associated with chromatin of an intermediate density, is meaningless as the DNA molecules in this system are either fully density labeled (one entire strand) or not at all.

bution of radiolabel in subsequent mitoses using autoradiography. Their conclusions were two-fold: (a) there is turnover of chromosomal protein (which we would expect to be primarily nonhistone protein) and (b) the residual protein after several generations was uniformly distributed among all the chromatids. This direct observation, while clouded by an inability to distinguish between histone and other nucleoproteins, is nonetheless clearly consistent with the conclusions we have drawn from these experiments.

This conclusion would appear to infer that histones recognize the DNA molecule as a whole, rather than a specific aspect of one or the other strands, and that at replication a pre-existing histone or group of histones has an equal chance of becoming connected with either of the two daughter DNA molecules. This does not appear to be an unreasonable conclusion in view of what is known of histones in their role as chromosomal structural proteins. They can generate similar structures with DNA from widely diverse organisms, though they themselves change but little. Furthermore, specific information encoded in histone molecules for the specific deposition of new histone appears unlikely, since a single mammalian protamine species can be correctly replaced by histones after fertilization.

The results do not exclude the possibility that histones are deposited in an highly organized manner at the replication fork, and that they subsequently randomize. A study of this possibility is currently underway.

Acknowledgments

We wish to thank Dr. Charles S. Swenson for the use of his Dupont Curve Analyzer, without which this work could not have been completed. We thank our colleagues within this laboratory for their advice and the pleasant environment in which to work. This study was supported by the USPHS Grants #Ca 10871 and GM 10871.

References

1.   D. M. Prescott and M. A. Bender. Exp. Cell Res. 29:430 (1963).

2.   R. Tsanev and G. Russev. Eur. J. Biochem. 43:257 (1974).

3.   R. Hancock. J. Mol. Biol. 48:357 (1970).

4. V. Jackson and R. Chalkley. Biochemistry 13:3952 (1974).

5. D. Brutlag, C. Schlehuber and J. Bonner. Biochemistry 8: 3214 (1969).

6. A. J. Varchavsk and G. P. Georgiev. Biochim. Biophys. Acta 281:669 (1972).

7. R. Balhorn, D. Oliver and R. Chalkley. Biochemistry 11: 1094 (1972).

8. R. Hancock. J. Mol. Biol. 86:649 (1974).

9. G. A. Bray. Anal. Biochem. 1:279 (1960).

ANALYSIS OF SPECIFIC PHOSPHORYLATION SITES IN LYSINE-RICH (H1) HISTONE: AN APPROACH TO THE DETERMINATION OF STRUCTURAL CHROMOSOMAL PROTEIN FUNCTIONS

Thomas A. Langan and Philip Hohmann*
Department of Pharmacology
University of Colorado Medical School
Denver, Colorado 80220

Abstract

The phosphorylation of lysine-rich histone in the non-growing cells of adult rat liver which takes place in response to hormonal or cyclic AMP administration occurs on a specific serine residue (serine 37) in the histone. In contrast, the phosphorylation which takes place in growing cells involves sites other than serine 37, and occurs to a large extent on threonine as well as serine residues. Distinct enzymes catalyzing the two types of phosphorylation have been isolated. Cyclic nucleotide stimulated and growth-associated lysine-rich histone phosphorylation was studied in cultured Reuber H-35 hepatoma cells. Four phosphopeptide spots were obtained from tryptic digest of purified lysine-rich histone by high voltage paper electrophoresis and thin layer chromatography. One of these contains the cAMP-stimulated phosphorylation site of lysine-rich histone (serine 37) previously demonstrated in adult rat liver. In H-35 cells, incorporation of ^{32}P-phosphate into this site is small in extent, not correlated with the rate of cell growth, and stimulated an average of 5-fold by dibutyryl-cAMP. The limited response to cAMP is compatible with a selective phosphorylation of a small fraction of F_1 histone molecules in response to hormone stimulation. The remaining phosphopeptides isolated from the H-35 cell lysine-rich histone represent phosphorylation sites which are not observed in non-growing tissue. Incorporation of ^{32}P-phosphate into these sites is massive in extent, strongly correlated with the growth rate of the cells, and unaffected by cyclic nucleotides. The finding that there are two types of F_1 histone phosphorylation which can be distinguished on the basis of the sites phosphorylated, the regulatory stimuli involved, the extent of phos-

---

* Present address: Biomedical Research Group, H-9, Los Alamos Scientific Laboratory, Los Alamos, New Mexico 87544.

phorylation in vivo, and the enzymes catalyzing the reactions indicated the existence of multiple functional roles for F_1 histone phosphorylation. It is proposed that multiple functions are implemented by specific conformational changes in the histone brought about by phosphorylation of the different sites, which in turn lead to changes in chromatin structure appropriate to the particular function. In this way F_1 histone phosphorylation, together with other histone modification reactions, may provide a general mechanism for bringing about the various structural changes required for the function and replication of chromatin.

In keeping with current concepts of gene regulation in eukaryotes, the great majority of papers given at this colloquium describe studies of the nonhistone chromosomal proteins. This reflects the obvious fact that histones do not have the diversity required for the recognition of specific nucleotide sequences in DNA, and therefore must have an essentially structural role in chromatin function. The primary regulatory components are taken to be chromosomal proteins other than histones, and ample evidence for the existence of nonhistone proteins which regulate specific genes is emerging from a number of laboratories. However, another fact emerging from current studies of nonhistone chromosomal proteins is that the great bulk (on a weight basis) of the nonhistone protein also has only the potential for some type of structural role in chromatin, since it exhibits only a moderate tissue and species specificity and is incapable of recognizing specific nucleotide sequences in DNA. Therefore, all but a fairly small fraction of total chromosomal protein appears to have what we can term loosely as an essentially structural role in chromatin. Presumably the highly conserved "core" histones[1] H4, H3, H2a and H2b are involved in maintaining the structural elements that form the common, elementary repeating units of all chromatin. H1 histone, being more diversified, may be involved in forming more varied (or less demanding) packing arrangements of the elementary repeating units, while much of the nonhistone protein may generate various specific structures required in, for example, different active regions of the chromatin. None of these structural functions would involve the direct recognition of DNA sequences, yet given the close relationship between structure and function generally found in biology, it would be surprising if these structures did not influence the transcription and replication of the genes in the DNA contained within them.

---

[1] The 1974 Ciba symposium histone nomenclature is used. This nomenclature corresponds to the Johns nomenclature (F_1, F_2a_2, F_2b, and F_3) with the exception that F_2a_1 = H4.

Thus, the chromosomal proteins which regulate gene expression can be thought of as falling into two broad functional classes, those that act as specific sequence-recognizing proteins and those that act to determine general structural features which affect the activity of chromatin. Both histones and nonhistones may participate in the latter process. Understanding the regulation of the genome then appears to involve understanding the function of both sequence-recognizing and structure-determining proteins. For investigating the action of sequence-recognizing proteins on the transcription of specific genes, the use of complementary DNA to assay specific mRNA products provides a promising approach, as evidenced by several papers presented elsewhere in this volume. Investigating the influence of more general structural features on chromatin activity is less straight-forward because of the difficulty of recognizing and distinguishing various structural states of chromatin. A property which is shared by many histones and nonhistone proteins which shows some promise as an approach to this problem is that these proteins are subject to protein modification reactions - in particular to phosphorylation - to a remarkable degree compared to cell proteins in general. Phosphorylation undoubtedly influences the structure of the proteins, and most likely the sturcture of the chromatin of which the proteins are a part. As discussed further below, changes in the phosphorylation state of chromosomal proteins may be the means by which changes in chromatin structure are brought about. In addition, examining the phosphorylation state of a particular chromosomal protein may provide a means for recognizing and detecting the existence of a particular structural state in chromatin, and perhaps for correlating this state with a particular activity of chromatin, whether this be changes in the transcription rate in a certain fraction of chromatin, or broader aspects of gene expression and regulation such as replication of the chromatin or the segregation of chromosomes at mitosis.

In this paper we will outline our work on the phosphorylation of one class of chromosomal proteins, the lysine-rich (H1) histones. The phosphorylation of H1 histone exhibits a rather impressive degree of variety, with various phosphorylation states correlating with specific stimuli or cellular activities, more or less in keeping with the ideas discussed above. Finally, this discussion will be extended by some further remarks on the role of protein phosphorylation in the functioning of histones and perhaps other structure-determining chromosomal protein.

115

Characterization of Phosphorylation Sites in Lysine-Rich Histone.

Several years ago we characterized two phosphorylation sites in lysine-rich histone, one occurring at serine 37 (1,2) and one at serine 106 (2,3) (Fig. 1). The location of these phosphorylation sites in the sequence of H1 histone has been established by comparing the sequence of tryptic phosphopeptides isolated from the histone with the overall sequence of H1 histone as it is being worked out by David Cole and his colleagues (4,5).

Each of these phosphorylation sites is phosphorylated by a specific enzyme, and these can be purified and separated from one another by conventional methods of protein purification (6, 2,3,7). The enzyme which catalyzes phosphorylation at serine 37 is identical with the ubiquitous cyclic AMP-dependent protein kinase which, we have been able to show, has H1 histone as one of its several physiological substrates (1,8).

Our interest in the characterization of phosphorylation sites in H1 histone stems not only from a desire to establish the relationship of these phosphorylation sites to other structural features in the histone molecule, such as DNA binding sites or regions of secondary structure, but also from the need that became apparent for a reliable and selective method for evaluating the different types of H1 histone phosphorylation which occur in cells. By the isolation of specific tryptic phosphopeptides from the histone, one can recognize and distinguish the phosphorylation of distinct sites in H1 histone, which, as shown below, are phosphorylated under specific conditions and as a result of specific stimuli.

As an example of this technique, Figure 2 shows an experiment characterizing the types of lysine-rich histone phosphorylation catalyzed by enzymes present in a crude extract of rat liver. Two phosphopeptides are resolved by high voltage paper electrophoresis of tryptic digests of the phosphorylated histone. Peptide A contains serine 37 and peptide B contains serine 106. In the absence of cyclic AMP, the phosphorylation of these two sites (catalyzed by two separate enzymes in the crude extract, as mentioned above) is approximately equal. In the presence of cyclic AMP, there is a large stimulation of the enzyme catalyzing phosphorylation at serine 37, but no change in the phosphorylation of serine 106. Thus the analysis of phosphopeptides containing specific phosphorylation sites gives considerably more information about the phosphorylation of H1 histone in these extracts than analyses which depend on the measurement of phosphate content or ^{32}P incorporation into

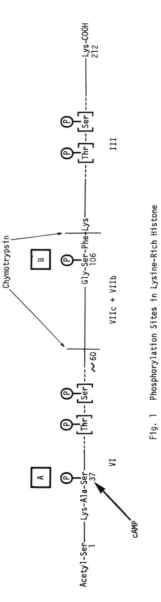

Fig. 1   Phosphorylation Sites in Lysine-Rich Histone

117

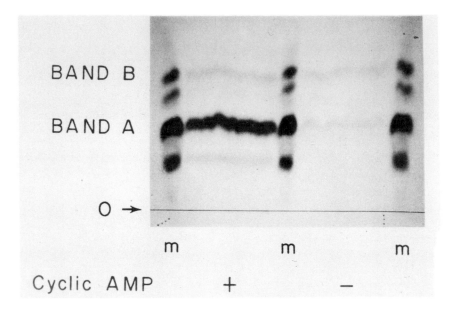

Fig. 2. Phosphorylation of H1 histone by a crude extract of rat liver. A 100,000 g supernatant fraction of rat liver was used as a source of enzymes. Phosphorylation and isolation of phosphopeptides by high-voltage paper electrophoresis was carried out as described previously (8) and the phosphopeptides visualized by radioautography. Cyclic AMP, 5×10^{-6} M, was present in the reaction mixtures where indicated. m designates marker phosphopeptide spots. 0 = origin. Anode at top.

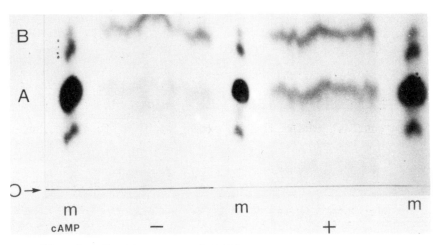

Fig. 3. Caption on next page.

118

the whole histone molecule.

Another example of this technique is shown in Figure 3, in which phosphopeptide analysis was used to characterize the histone kinase activity bound to washed chromatin preparations. It had been reported (9,10) that the histone kinase activity of chromatin shows little or no stimulation by cyclic AMP. The phosphopeptide analysis, however, shows clearly that phosphorylation of serine 37 is catalyzed by washed preparations of chromatin, and that this is stimulated substantially by cyclic AMP. However, a considerable amount of phosphorylation at serine 106 (together with the phosphorylation of endogenous nonhistone protein substrates by kinases in the chromatin), provides a high background of cyclic AMP-independent phosphorylation which obscures the stimulation of the cyclic AMP-dependent activity when total incorporation into TCA insoluble protein is measured. Therefore, the phosphopeptide analysis serves not only to distinguish various types of H1 histone phosphorylation, but also to separate these from other protein phosphorylation reactions occurring in various components or contaminants of the preparation. Further purification of specific H1 phosphopeptides can be, and is, employed when necessary to separate them from other phosphopeptides present.

By the use of phosphopeptide analysis, the phosphorylation of residue 37 of H1 histone by cAMP-dependent protein kinase in adult rat liver in response to administration of glucagon and cyclic AMP has been demonstrated (1,8,11). One interesting feature of this hormonally regulated histone phosphorylation is that it is limited to a very small fraction of total H1 histone molecules (estimated at about 1% in rat liver) even though isolated histone can be phosphorylated stoichiometrically at serine 37 by the purified enzyme. The basis for the limited phosphorylation in vivo is not known, but it is clearly in keeping with some type of selective response of a small fraction of the chromatin to hormonal stimulation.

Growth-Associated Lysine-Rich Histone Phosphorylation

In contrast to the limited extent of H1 histone phosphorylation which occurs upon hormonal stimulation of the nongrowing cells of adult rat liver, massive phosphorylation of H1 histone, involving essentially all H1 molecules, occurs in rapidly growing cells in association with both S phase and mitosis

Fig. 3. Phosphorylation of H1 histone by washed rat liver chromatin. An extensively washed chromatin preparation was used as a source of enzymes. Other conditions as in Figure 2. m designates marker phosphopeptide spots. 0 = origin.

(12-16). Investigation of the histone kinase activity present in rapidly growing cells showed that these cells contain an additional chromatin-bound histone kinase activity which catalyzes the phosphorylation of specific sites in H1 histone, distinct from serines 37 and 106 (17). In contrast to other known histone and protein kinases, the phosphorylation catalyzed by the growth-associated histone kinase occurs to a greater extent on threonine than on serine residues (17). Extensive phosphorylation of threonine residues also occurs in vivo in H1 histone of rapidly growing Ehrlich ascites and Reuber hepatoma cells (17). Bradbury et al. (16) also observed a variable but occasionally substantial proportion of threonine phosphorylation in H1 histone of dividing cultures of the slime mold Physarum polycephalum. The isolated enzyme activity is independent of cyclic AMP and appears to be similar or identical to that observed by Lake and Salzman in HeLa and other growing cells (18,19). The growth-associated histone kinase has been purified and separated from the enzymes catalyzing the phosphorylation of serines 37 and 106, thus making a total of three distinct histone kinases which can be isolated from tissues or cells. The nature of the phosphorylation sites in H1 histone phosphorylated by the growth-associated histone kinase is under investigation. We have established that they occur exclusively in or near the basic DNA binding regions of H1 histone in the N and C terminal ends of the molecule, and are absent from the central, neutral-hydrophobic region (17). There is at least one threonine and one serine-containing phosphorylation site at each end of the molecule (Figure 1), although all sites may not be present in all subfractions of H1 histone.

Growth-Associated and Hormone-Regulated H1 Histone Phosphorylation in Cultured Cells

We have utilized the method of phosphopeptide analysis to investigate and compare the types of H1 histone phosphorylation which occur in growing and non-growing cells in the presence and absence of stimulation by cyclic AMP (20). For these experiments we used cultures of the cyclic AMP-sensitive Reuber H-35 hepatoma cell line. The phosphopeptide analysis revealed that tryptic digests of H1 histone from growing cells contain small amounts of the phosphopeptide containing serine 37, together with larger amounts of a minimum of three phosphopeptides corresponding closely to the phosphopeptides isolated from H1 histone phosphorylated by purified growth-associated histone kinase preparations.

In order to determine the effect of cell growth on the phosphorylation of different sites in H1 histone, the growth of H-35 hepatoma cells was altered by withholding serum from the medium. As a measure of the rate of cell growth, the incorporation of ^{3}H-(methyl)-thymidine or ^{3}H-deoxyadenosine into DNA was measured.

The phosphorylation of various sites in H1 histone during the decline in growth rate which follows a change to serum-free medium is shown in Table I. Phosphorylation of the sites contained in peptide spots A3 plus A4 decreases with decrease in

TABLE I

Phosphorylation of Various Sites in H1 Histone During Periods of Decreasing Cell Growth

Day	CPM ^{32}P in phosphopeptides per 100 μg DNA		CPM ^{3}H-Thymidine per μg DNA
	peptide spots A3 + A4	peptide A1	
0	2000	96	880
1	2308	87	910
2	-	-	450
3	740	88	100
4	288	48	95

Serum was withdrawn from the medium 36 hours after subculture (Day 0). Cells were exposed to ^{32}P-phosphate for two hours at the times indicated following serum removal. ^{3}H-thymidine incorporation was measured during a one hour period in separate flasks. H1 histone phosphopeptides were isolated and purified as described by Hohmann and Langan (20).

cell growth, while phosphorylation of the site in peptide A1, containing serine 37, is virtually unchanged. The phosphorylation of various sites in lysine-rich histone as growth resumes after cells maintained in serum-free medium are subcultured into medium containing serum is currently being studied.

The effect of dibutyryl cyclic AMP on H1 phosphorylation was also studied in cultured H-35 cells. Cells were maintained in serum-free medium until growth had essentially ceased. As noted above, such cells are fully viable and resume growth when cultured in the presence of serum. Addition of dibutyryl cyclic AMP to these cells causes a four to six-fold increase in phosphorylation of serine 37 (peptide A1) while it has no effect on the small residual phosphorylation taking place at the growth-associated phosphorylation sites in the peptides of spots A3

121

plus A4 (Table II). This observation shows that phosphorylation of the growth-associated sites is independent of cyclic

TABLE II

Effect of Dibutyryl Cyclic AMP on the Phosphorylation of
Various Sites in H1 Histone

	CPM ^{32}P in phosphopeptides	
Dibutyryl cyclic AMP	Peptide A1	Peptide spots A3 + A4
-	17	92
+	110	100

Cells maintained in serum-free medium for four days were
exposed to ^{32}P-phosphate for two hours in the presence or absence of dibutyryl cyclic AMP (0.5 mM). Peptide isolation as
in Table I.

AMP stimulation in vivo. It also shows that the increased ^{32}P-
phosphate incorporation into serine 37 in the presence of dibutyryl cyclic AMP is not due to an increased transport of radioactive phosphate into the cell, resulting in a higher specific
activity of ATP pools. In similar experiments, it was shown
that phosphorylation of serine 37 is stimulated by cyclic AMP
to a similar degree in both growing and non-growing cells.
Thus, independent regulation of phosphorylation at growth-associated and cyclic AMP-sensitive phosphorylation sites occurs
under all the conditions studied.

DISCUSSION

The studies described above establish that H1 histone is
subject to multiple types of specific modification by phosphorylation. Histones, as noted above, appear to function essentially as structure-determining proteins in chromatin. However,
the structure of chromatin is dynamic rather than static, and
changes in this structure undoubtedly influence transcriptional
activity. In addition, structural changes are clearly required
during mitosis and chromatin replication. It has been suggested that the phosphorylation of H1 histone may produce conformational changes in the histone, and thus provide a mechanism for
bringing about required structural changes in chromatin (21).
Consistent with this possibility, physical studies have established that phosphorylation of H1 histone can substantially
alter its interaction with DNA (22,23). Also, chromatin recon-

stituted with phosphorylated Hl histone is a better template for RNA synthesis than chromatin reconstituted with unphosphorylated histone (24). It is now a common assumption that histone modifications can serve to alter the interaction between histones and DNA or other elements of chromatin.

The finding that there are two types of Hl histone phosphorylation which can be distinguished on the basis of the sites phosphorylated, the regulatory stimuli involved, the extent of phosphorylation in vivo, and the enzymes catalyzing the reactions suggests further that multiple functional roles for Hl histone phosphorylation may exist. These multiple functions may be implemented by specific conformational changes in the histone brought about by phosphorylation of specific sites, which in turn lead to changes in the structural state of the chromatin appropriate to the particular function.

If in fact the phosphorylation of specific sites is associated with specific structural states in chromatin, one might expect the phosphorylation of Hl histone which occurs in mitosis to take place on different sites than those phosphorylated in S phase, since the structure of mitotic and S phase chromatin is clearly different. Hohmann, Tobey and Gurley (25) have investigated this possibility in synchronized cultures of Chinese hamster ovary cells, and found that Hl histone phosphorylation in mitotic cells occurs at one or more threonine-containing phosphorylation sites which are not detectable in randomly growing cells (in which almost all phosphorylation would represent S phase phosphorylation). This finding gives considerable support to the idea that the phosphorylation of different specific sites in Hl histone can bring about different structural changes in chromatin, allowing Hl histone to function in multiple ways in the determination of chromatin structure.

The fact that phosphorylation of chromosomal proteins is not limited to Hl histone, but includes other histones as well as nonhistone chromosomal proteins, together with the existence of at least one other type of reversible histone modification, acetylation, suggests that the phosphorylation of Hl histone revealed in these studies may be just one example of a quite general mechanism for bringing about various structural changes required for the function and replication of chromatin. If this is the case, study of the state of modification of chromosomal proteins under various conditions of growth and function may provide a method for the recognition of various structural states in chromatin, and provide a means for evaluating the contribution of proteins that determine chromatin structure to the regulation of gene expression and other chromosomal functions.

Acknowledgement

We thank Dr. W. D. Wicks for generously supplying cultures of Reuber H-35 hepatoma cells, and for advice on the growth of this cell line. This work was supported by the United States Public Health Service, Grant CA 12877 from the National Cancer Institute. P.H. was supported by a Postdoctoral Fellowship from the National Cancer Institute.

References

1. T. A. Langan. Proc. Nat. Acad. Sci. 64:1276 (1969).

2. T. A. Langan. Ann. N. Y. Acad. Sci. 185:166 (1971).

3. T. A. Langan. Fed. Proc. 30:1089Abs (1971).

4. S. C. Rall and R. D. Cole. J. Biol. Chem. 246:7175 (1971).

5. G. M. T. Jones, S. C. Rall and R. D. Cole. J. Biol. Chem. 249:2548 (1974).

6. T. A. Langan. Science 162:579 (1968).

7. J. Vandepeute and T. A. Langan, in preparation (1975).

8. T. A. Langan. J. Biol. Chem. 244:5763 (1969).

9. G. Siebert, M. G. Ord and L. A. Stocken. Biochem. J. 122: 721 (1971)

10. R. H. Burdon and C. A. Pearce. Biochim. Biophys. Acta 246:561 (1971).

11. L. E. Mallette, M. Neblett, J. H. Exton and T. A. Langan. J. Biol. Chem. 248:6289 (1973).

12. R. Balhorn, R. Chalkley and D. Granner. Biochemistry 11: 1094 (1972).

13. A. J. Louie and G. H. Dixon. J. Biol. Chem. 247:5498 (1972).

14. D. B. Marks, W. K. Paik and T. W. Borun. J. Biol. Chem. 248:5660 (1973).

15. L. R. Gurley, R. A. Walters and R. A. Tobey. J. Cell Biol. 60:356 (1974)

16. E. M. Bradbury, R. J. Inglis, H. R. Matthews and N. Sarner Eur. J. Biochem. 33:131 (1973).

17. T. A. Langan and P. Hohmann. Fed. Proc. 33:1597 (1974).

18. R. S. Lake and N. P. Salzman. Biochemistry 11:4817 (1972).

19. R. S. Lake. J. Cell Biol. 58:317 (1973).

20. P. Hohmann and T. A. Langan. (manuscript in preparation).

21. T. A. Langan. In: "Regulatory Mechanisms for Protein Synthesis in Mammalian Cells", A. San Pietro, M. R. Lamborg and F. T. Kenny, eds., p. 101, Academic Press, New York (1968).

22. A. J. Adler, B. Schaffhausen, T. A. Langan and G. D. Fasman. Biochemistry 10:909 (1971).

23. A. J. Adler, T. A. Langan and G. D. Fasman. Arch. Biochem. Biophys. 153:769 (1972).

24. G. Watson and T. A. Langan. Fed. Proc. 32:588Abs (1973).

25. P. Hohmann, R. A. Tobey and L. R. Gurley. Biochem. Biophys. Res. Commun. 63:126 (1975).

Note Added in Proof:

Although technical problems have not been completely ruled out, recent experiments indicate that there is a significant increase in phosphorylation in the cyclic AMP-sensitive phosphorylation site when cells are caused to resume growth. While this increase is small in absolute amount, it is proportional to the increases which occur in the major growth-associated phosphorylation sites. Therefore, in view of the data available at this time, the conclusions drawn above that cyclic AMP-stimulated phosphorylation and growth-associated phosphorylation reactions are completely distinct functionally must be modified.

HISTONE METHYLATION, A FUNCTIONAL ENIGMA

Paul Byvoet and C. Stuart Baxter
Department of Pathology
University of Florida
College of Medicine
Gainesville, Florida 32610

Abstract

Methylation of basic amino acid side chains in histones occurs "in situ" in chromatin, mainly on specific internal lysine residues in histone F_3 (nos. 9,27) and F_{2a1} (no. 20). Although methylation of arginine, histidine and carboxyl side chains has been reported, these processes constitute only a minor proportion of the total in vivo incorporation of radio-methyl into histones after administration of methionine-Me-^{3}H. The methylated lysine residues in F_3 and F_{2a1} are situated in the polar regions of these histone molecules, and since methylation increases hydrophobicity and cationic charge of the amino groups, it seems likely that this process increases the binding of these polar regions to DNA. Aside from the fact that F_3 and F_{2a1} generally behave as the most cationic histones, (Biorex chromatography, displacement from chromatin) studies in this laboratory have attempted to correlate increased binding and methylation of histones in chromatin. 1) Gradual displacement of radiomethylated histones isolated from nuclei preincubated in the presence of S-adenosylmethionine-Me-^{3}H (SAM-^{3}H) with increasing concentrations of HCl, protamine, polylysine and deoxycholate indicate that those histones released at the highest concentrations, were labeled to the greatest extent. 2) Partial trypsinization of chromatin isolated from liver nuclei incubated in the presence of SAM-^{3}H removes no acid-extractable label from chromatin. This type of data suggests a more tenacious binding of either methylated as opposed to nonmethylated histones in chromatin, in general, or more specifically, of newly methylated histones in premitotic chromatin, since methylation occurs only in G_2 just prior to mitosis. Studies on the influence of foreign substances on histone lysine methylation in nuclei incubated in the presence of SAM-^{3}H, indicate that activated carcinogens have a general tendency to inhibit, whereas intercalating agents and polylysines tend to stimulate this process to various degrees. Structural analogs of intercalating agents unable to intercalate because of bulky substituents were inhibitory or without effect, while none of the above mentioned agents stimulated the methylation of free histone in solution by the isolated histone lysine-methyltransferase in the presence of SAM-^{3}H. Stimulation by polylysines depends on their degree of polymerization, but always reaches a maximum at 50% saturation of DNA-P with lysine residues. Competition experiments with intercalating agents of known affinity for either large or small groove suggest that in order to stimulate, polylysines have to bind to the large groove, and that this only occurs with polymers containing more than 9 lysine residues per molecule. The marked effects resulting from relatively small structural changes in DNA suggest a close proximity of DNA, polar histone region and histone methyltransferase.

Since the presence of methyllysine in histones was discovered by Murray in 1964 (1), the ensuing decade has seen the development of a large amount of detailed information on the process of histone methylation (2). The biological significance of this postsynthetic histone modification, however, still remains a complete mystery.

Histone methylation is an enzymatic process and occurs on amino and carboxy-groups utilizing S-adenosylmethionine as methyl donor. After in vivo administration of methyllabeled methionine, the major portion of radiomethyl is incorporated into the ε, N-methyllysines (mono-, di- and tri-)(3,4). Other methylated amino acid derivatives such as methylarginine (5), methylhistidine (6) and carboxymethylated (2) amino acids have been reported but seem to occur in much smaller amounts. Studies on the primary structure of histones, for example, have so far not indicated the presence of any methylated amino acids other than methyllysine (7). Data from our laboratory (8) on the in vivo incorporation of radiomethyl from methyllabeled methionine into histones from rat tissues are shown in Table I. It is

TABLE I

Relative Radioactivity Recovered in Methylated Amino Acids of Histones from Rat Tissues

Tissue	N-methyl-lysines	N-methyl-histidine	N-methyl-arginine
Liver	1.00	0.106±0.021	0.332±0.050
Thymus	1.00	0.044±0.011	0.032±0.018
Novikoff hepatoma	1.00	0.126±0.014	0.130±0.017

Caption on next page.

evident that compared to the incorporation into methyllysine, that into methylhistidine and methylarginine was rather small in all three tissues studied. The peculiar constancy of the relative distribution of label in these tissues over a long time interval would seem to indicate some degree of tissue specificity of the methylation pattern.

The above-mentioned studies on histone primary structure have also indicated that the only methyllysine-containing histones are F_{2a1} and F_3, and that these residues are situated at positions 20 in F_{2a1} and 9 and 27 in F_3 (7). Interestingly, the methylated lysine residues are always situated 3 residues away from an acetylated (or "acetylatable") lysine. This distance happens to be just one turn in an α-helix. Most studies on free histones in solution indicate however that the polar parts of the histone molecules do not contain α-helix, although this does not rule out the possibility that it may exist in the native chromatin complex.

A number of studies using sources other than calf thymus have indicated that F_{2b} may also be methylated (9,10). Of course, it should be kept in mind that most sequence studies which located methylated residues have been done on calf thymus histones and that the conservatism seen in the primary histone structure is not reflected in the methylation pattern, as is demonstrated by the data in Table I and by the fact that pea seedling histone F_{2a1} is not methylated at all (7).

During extensive studies on protein methylating enzymes, Paik and Kim have succeeded in extracting, purifying and characterising three methyl-transferases with specificity for arginine (11), carboxyl-groups (12) and lysine (13), which they called methylase I, II and III, respectively. The first two were found mainly in the cytosol, in contrast to the lysine methyl-transferase activity which is primarily nuclear. In agreement with this, we have never found any activity other than that of lysine methyl-transferase in isolated nuclei. It appears therefore that this enzyme which is responsible for the major portion of in vivo histone methylation, is the only

---

Table I. Hepatoma-bearing rats received 500μCi/kg methionine-(Me-^{14}C) and were killed 2, 24, and 48 hours later. Whole histones minus F_1 were isolated, hydrolyzed and analyzed for basic amino acids (8). Methyllysines were not separated. Total counts in the various radioactive peaks were determined (8) and ratios calculated on the basis of methyllysines being unity. Data are averages of 6 determinations and their standard errors.

129

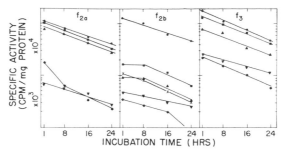

Methionine (●); total methyl lysines (■); monomethyl lysine (◄►); dimethyl lysine (▲); trimethyl lysine (▼) and methyl arginine (×).

Fig. 1. Turnover of histone fractions and their methyl groups from CHO-cells. Chinese hamster ovary cells were grown in the presence of methionine-Me-^{14}C for several hours. Cells were harvested and resuspended in cold medium and aliquots were harvested at appropriate time intervals. Histone fractions were isolated from isolated nuclei and specific activities determined with respect to methionine-^{14}C and radiomethylated lysine or arginine (10). Curves are semi-log linear regressions of specific activities of F_{2a1}, F_{2b} and F_3.

methyltransferase which is firmly bound within the nucleus and in all probability an integral part of chromatin. Since methyl-arginine, methylhistidine and carboxymethyl groups have been reported in histones (2,8) it would seem that at least small amounts of the other methyltransferases are present in the nucleus but leak out at some point during isolation. Studies in our laboratory indicate that the purified lysine or arginine methyltransferase will not methylate any histones in chromatin incubated in the presence of these enzymes and methyllabeled S-adenosylmethionine.

Studies by Tidwell (14) and Shepherd (9) have shown that histone methylation takes place exclusively at the end of S-phase, extending into G_2, and it is therefore tempting to speculate that it might be involved in preparation of chromatin for mitosis. This would seem to be unlikely, since this histone modification appears to be irreversible. In order to study the turnover of methyl groups relative to the histone peptide backbone, CHO-cell cultures were grown in the presence of methionine-Me-^{14}C for several hours. The cells were then harvested and resuspended in cold medium, and aliquots harvested at appropriate time intervals to determine the decay of the specific radioactivity of histones with respect to radiomethyl and to compare that with the decay of the specific radioactivity of the peptide backbone with respect to methionine-Me-^{14}C.

The results of these experiments, shown in Figure 1, indicate very clearly that the decay of the specific activities with respect to methylated derivatives and to methionine were approximately equal, as indicated by the parallelity of their linear regressions. These results imply that there is no turnover of methylgroups relative to the peptide backbone (10). It is interesting to note that F_{2b} was the only histone fraction containing labeled methylarginine.

The absence of turnover was subsequently confirmed in vivo in a number of rat tissues. Using the same approach, methyl-labeled methionine was administered to rats and the animals killed 3 and 50 hours later. Arginine-rich histones were isolated and specific radioactivities determined with respect to methionine, methyllysine and methylarginine. The results summarized in Table II indicate that in vivo the decay rates of the various methylgroups were similar to that of the peptide backbone (15).

TABLE II

Relative Distribution of Radioactivity Over Methylated Basic Amino Acids of Histones From Rat Tissues

Tissue	Time after injection of ^{14}C-methionine (hour)	Methionine	ε-N-methyllysines				ω-N-methyl arginine
			Total	Mono-	Di-	Tri-	
Spleen	3	1.00	0.435				0.023
	50	1.00	0.448				0.035
Liver	3	1.00	0.802	0.282	0.412	0.126	0.096
	50	1.00	0.662	0.142	0.401	0.128	0.113
Novikoff	3	1.00	0.409	0.122	0.283	0.030	0.075
hepatoma	50	1.00	0.392	0.043	0.348	0.018	0.101

Hepatoma bearing rats received lmCi/kg methionine-(Me-^{14}C) and were killed 3 and 50 hours later. Whole histones minus F_1 were isolated, hydrolyzed and analyzed for basic amino acids (15). Total cpm in the radioactive peaks were determined and ratios calculated on the basis of methionine being unity. Data are averages of 2-3 separate animal experiments. Average cpm in the methionine fraction at 3 and 50 hour in spleen, liver and tumor were respectively: 14,400; 5360; 1852; 1338; 8456; 5817.

Considering the timing of methylation of histones in the cell cycle, and their metabolic stability (16), one could envision that it may be involved in some finalization process, which follows completion of the synthesis of histones and their initial integration into the chromatin structure. We hypothesize that the methylation process would serve to more firmly lock the histones in place by increasing their binding to other macromolecules, to form a stable chromatin complex, as for instance the 11S particles proposed by Noll (17) or ν-bodies proposed by the Olins (18). (Footnote 1)

Although there is no doubt that methylation increases the hydrophobicity of thus modified amino-groups (19), some argument exists concerning the increase in basic character.

Paik and Kim (2) have argued that whereas monomethylation of protein lysine residues would increase their basicity, dimethylation would have the reverse effect. In the absence of other factors, therefore, ionic DNA-histone interactions would be increased in strength by monomethylation, but decreased by dimethylation of histone lysine residues.

The "anomalous" basicity order of simple aliphatic amines in aqueous solution $(-N(CH_3)_2 < -NH_2 < -NH CH_3)$ must be ascribed to solvation effects (20,21). Due to their high pKa values, the ε-amino functions of lysine and methyllysines are virtually completely protonated at physiological or lower pH, hence a more accurate measure of the relative affinities of the latter amines for DNA phosphate anions is the charge density at the nitrogen and next-neighbor carbon atoms. Relative changes in this quantity may be estimated by carbon-13 (CMR) spectroscopy, since within a series of structurally related compounds they are found to correlate with corresponding changes in carbon-13 NMR chemical shift at the same atom (22).

The CMR spectra of lysine and ε-N-methylated analogues in water at pH 5 (Figure 2) show that the chemical shifts of the ε- and N-methyl carbon atoms increase isotonically, and almost additively, with increasing methyl substitution at lysine, and provide good evidence for a regularly increasing positive charge in the region of the ε-amino function.

---

1. We would like to point out that an important factor in the concept of a stable, elementary chromatin particle, is the knowledge that histones turn over at the same rate as the DNA with which they are complexed in chromatin.

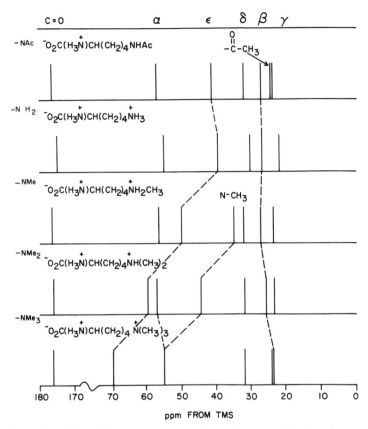

Fig. 2.   Graphic representation of natural-abundance
Fourier Transform Carbon -13 NMR spectra of lysine, ε-N-acetyl-
and ε-N-methylated lysines in water, (50 mg/ml) at 40° and pH 5.
Spectra were recorded on a Bruker HFX-90 spectrometer at 22·63
H$_z$ with proton broadband decoupling.   Each spectrum recorded
was the sum of 4,000 accumulations.   TMS=Tetramethylsilane

Since these observations suggest that lysine: DNA-phosphate binding in DNA-histone complexes increases regularly with
increasing lysine methylation, attempts were made to correlate
histone binding and methylation experimentally.

HISTONE DISPLACEMENT FROM CHROMATIN

Displacement of histones from chromatin has been studied by several laboratories, using HCl (23), NaCl (24), protamine (25) and deoxycholate (26), and the results generally indicate that the concentrations of a given displacing agent at which the particular histone fractions are released differ, although they overlap partially. Our aim was to repeat these experiments using chromatin isolated from nuclei incubated in vitro in the presence of S-adenosylmethionine-Me-^{3}H. In general, the chromatin was exposed to increasing concentrations of the displacing agent, followed by a final extraction with 0.25 N HCl to remove any remaining histone. Between extractions the chromatin was sedimented at 20,000 x g for 15 minutes. The data obtained in a representative experiment are shown in Table III.

TABLE III

Specific Activities of Histones Displaced with Increasing Concentrations of Polylysine* (MW 3000) from Rat Liver Chromatin Prelabeled in the Presence of S-adenosylmethionine (Me-^{14}C).

polylysine concn. in 0.3 M NaCl (mg/ml)	molar ratio lysine/DNA·P	histone displaced (mg)	specific activity (cpm/mg)
1	.15	1.57	732
5	.75	4.93	3,380
10	1.5	5.67	5,020
20	3.0	4.24	6,380
0.25 M HCl	0	5.11	10,100

*dodecalysine

Nuclei from 16 grams of rat liver were incubated at 37° with 100 μCi SAM-methyl^{3}H in 0.01 M Tris/HCl, 0.25 M sucrose, 0.025 M KCl, 0.01 M MgCl$_2$, pH 8.9, total vol. 5 ml. After 10 minutes nuclei were chilled and washed twice with pH 7.2 saline-EDTA (0.08-0.02M) (49) and once with 0.3 M NaCl, 10mM MES buffer, pH 6.3 (28) (centrifugation at 2000 g for 20 minutes). To the isolated chromatin polylysine was added in the concentrations indicated, with a total volume of 3 ml containing 0.3 M NaCl and 10 mM MES buffer, pH 6.3. Between extractions chromatin was sedimented at 20,000 g for 15 minutes. Histone concentration was determined at 276 nm, at which polylysine has no absorption. All values are averages of duplicate determinations. In other displacement experiments mentioned in the text, protein concentrations were determined by fluorometry (29). (continued on the next page)

It can be seen that there was a continuous increase in specific activity of the displaced histones with increasing concentration of displacing agent, indicating that those histone species bound most tenaciously were labeled to the greatest extent. Essentially similar data were obtained with HCl, protamine and deoxycholate (27).

PARTIAL DIGESTION OF CHROMATIN WITH TRYPSIN

This experiment was fashioned after the original studies of Simpson (31), who reported that trypsin digestion removed 30-55% of all chromosomal proteins, and that all histone fractions were susceptible to it. His data suggest that the firmly bound regions of the chromosomal proteins were protected from the digestive action. We have repeated this experiment, using chromatin from nuclei incubated in vitro with S-adenosylmethionine-Me-^{3}H. The results, summarized in Table IV, show clearly

TABLE IV

Trypsinization of Chromatin Incubated
with S-adenosylmethionine-methyl-^{3}H

Treatment	Fraction	Protein mg/mg chromatin DNA	Total cpm/mg chromatin DNA
Control	pH 4.6 acetate solubilized	--	1,300
	HCl extract of residue	1.07	7,200
Trypsinization	Trypsin solubilized	0.78	10,540
	HCl extract of residue	0.62	7,950

Caption to Table IV on next page.

Table III - continued caption: Histones and protamines were separated by chromatography on Bio-Gel P-10. Exposure of chromatin to polycations appears to yield uncontaminated histones (30).

135

that the trypsin digestion did not remove any acid-extractable counts from the chromatin. This implies that although histones are partially digested, those portions containing the methyl label remain bound to DNA. Inferring from the sequence studies on F_{2a1} and F_3 from calf thymus (7), the methylated lysine residues are located in the polar portions of rat F_{2a1} and F_3, which are indeed the regions generally assumed to be firmly bound to DNA (32).

These data suggest that the methylated histones are bound more tenaciously to the DNA than other species and that this binding involves the polar regions. It is as yet not possible to conclude that methylated F_{2a1} or F_3 would bind better than the unmethylated versions. In addition, since methylation occurs exclusively at the very end of S-phase, moving into G_2, it can be concluded that those histones which incorporate radiomethyl are contained in cells preparing for mitosis. Their binding characteristics may therefore be a special structural feature of premitotic and mitotic chromatin.

INFLUENCE OF FOREIGN SUBSTANCES ON HISTONE METHYLATION

Carcinogens.

Since it is often possible to gain insight into certain processes by observing the result of their inhibition, we decided to search for compounds that would inhibit histone methylation by histone methyltransferases. In these studies, lysine methyltransferase activity was measured by incubating nuclei in vitro in the presence of S-adenosylmethionine-Me-^{3}H, as described earlier (33). In order to study the arginine methyltransferase, the enzyme was isolated according to Paik and Kim (11) and incubated in the presence of whole histone and S-adenosylmethionine-Me-^{3}H (34).

The results of these experiments, summarized in Table V

Table IV. Nuclei from 10 g rat liver were incubated with 100 μCi SAM-methyl-^{3}H and chromatin isolated as described in Table III. The chromatin pellet was suspended in the Tris/HCl buffer (DNA concentration; 1 mg/ml), and an aliquot vigorously stirred at 25° with trypsin (2% of total protein) added as a 1% solution in 1mM HCl for 25 minutes. Digestion was stopped by adjusting to pH 4.6 with the same volume of 0.2 M sodium acetate buffer, and centrifugation at 110,000 g (5 hr.). Another aliquot was treated in the same way, omitting trypsin. In each case final sediments were extracted with 0.24 N HCl, and extracts assayed for radioactivity.

TABLE V

Effect of SAM Analogues, Carcinogens and tRNA Methyltransferase Inhibitors on Histone Lysine and Arginine Methyltransferases

Type of Agent	Compound	% inhibition of methylation	
		lysine	arginine
SAM analogue	S-adenosylethionine	($K_I = 1.7 \times 10^{-4}M$, competitive)	($K_I = 10^{-5}M$, noncompetitive)
	S-adenosylhomocysteine	($K = 1.2 \times 10^{-5}M$, competitive)	($K_I = 5 \times 10^{-6}$, "mixed" inhibition kinetics)
Carcinogen*	N-hydroxy-2-AF	57	72
	N-hydroxy-2-AAF	12	29
	N-acetoxy-2-AAF	32	69
	Dimethylnitrosamine	0	0
	MNNG	50	33
	L-Ethionine	0	0
	9, 10-DMBA	0	0
	3'-methyl DAB	0	0
t-RNA methyl-* transferase inhibitor	Adenine sulfate	0	16
	N-6-Isopentenyladenosine	0	79
	Tubercidin	0	72

* Concentration = 1.0 mM.

Caption to Table V on next page.

show that both processes were inhibited by S-adenosylhomocys-teine, (a reaction product) and S-adenosylethionine, (a carcin-ogen metabolite). In contrast to the lysine methyltransferase, the arginine methyltransferase utilized both alkyl donors S-adenosyl-methionine and -ethionine indiscriminately, which ex-plains an earlier report (35) concerning uptake of radioactiv-ity into arginine residues of rat liver histone following in vivo administration of ethionine[14]C.

Both reactions were also inhibited by the proximal carcin-ogen N-acetoxy-2-acetylamino-fluorene and the alkylating agent MNNG, whereas the tRNA methyltransferase inhibitors only inhib-ited the arginine methyltransferase. These data suggest that at least several carcinogenic agents may interfere with histone methylation in vivo. Since histones do not turn over (16) and methylation is irreversible (9), it may be assumed that the re-sulting aberrant methylation pattern is permanent. It is at present not possible to conclude that this may play a role in the carcinogenic process, since carcinogens have been found to affect a number of other metabolic processes, as for example, stimulation of chromatin template activity and phosphorylation of nonhistone proteins (36).

Intercalating Agents.

Certain polar aromatic compounds are able to intercalate between the base pairs of DNA, with concomitant partial unwind-ing and lengthening of the double helix (37,38). The latter distortions of DNA would be expected to modify its interaction with polar, methylatable histone sequences, and hence the ac-tivity of the histone lysine methyltransferase in isolated nuclei.

Table V. Lysine methylation in nuclei from 0.5 g of rat liver was assayed as described in the subscript to Table III with 2.5 µCi SAM-Me-^{3}H in a total volume of 0.5 ml. Histones were extracted with 0.25 N HCl, precipitated in 25% TCA, the precipitate washed with 1% HCl in acetone, ether, dried and dissolved in water, and counted by liquid scintillation count-ing (33). Reaction mixture for histone arginine methylation contained: histone F_3 (0.5 mg), SAM-methyl-^{3}H (2.5 µCi, 2.5 nmoles) 0.5 M phosphate buffer, pH 7.2 (0.1 ml) and 0.1 ml methyltransferase preparation (13) in a total volume of 0.5 ml. Incubation was for 30 minutes at 37°, terminated by chilling and addition of 2.5 ml 18% trichloracetic acid. Purified his-tones were dissolved in water for liquid scintillation count-ing (34). Kinetic parameters and inhibition values deviated by ± 5% from the mean between different experiments.

TABLE VI

Effect of Polar Aromatic Compounds
on Histone Methylation in Nuclei

Intercalating	Maximal Stimulation (% of control)	n*
Acridine Orange	22	0.011
Actinomycin D	56	0.025(0.027)[39]
Dinitroaniline Reporter	22	0.170(0.250)[40]
Ethidium Bromide	72	0.036(0.033)[41]
Hycanthone	35	n.d.
Miracil D	40	0.125
9-chloro-MMD	11	n.d.
Naphthalene Bisimide Reporter	124	0.050
Naphthalimide Reporter	129	0.078
2, 9-Dimethyl NMP	38	n.d.
4, 7-Dimethyl NMP	45	0.20
2, 5, 6, 9-Tetramethyl NMP	54	n.d.
Proflavin	72	0.060

Non-intercalating

N'-methyl Miracil D (MMD)	-45**	--
4, 7-diphenyl NMP	0	--
3, 6-di-t-butyl Proflavin	0	--
Bis-N-methylacridinium nitrate	-18	--

* Maximum number of binding sites per nucleotide in DNA. Values in parentheses are those reported in the literature. n.d. = not determined.
** Negative sign denotes <u>inhibition</u> of methylation.

Nuclei were incubated with SAM-methyl-^{3}H, and histones extracted and assayed for radioactivity as described in the legend to Table 5. Agents were added as neutral chloride salts unless otherwise stated.

Surprisingly, such intercalating agents appeared to stim-ulate the incorporation of radiomethyl into histones from rat liver nuclei incubated in the presence of labeled S-adenosyl-methionine, without a simultaneous increase in histone extrac-ted (Table VI). The stimulation of methylation was observed

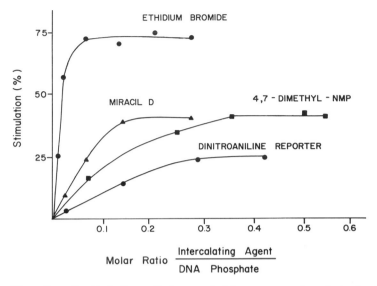

Fig. 3. Nuclei from 0.5 g rat liver were incubated with 2.5 μCi SAM—methyl-^{3}H for 15 minutes at 37° as described in the caption to Table III. Histones were extracted and assayed for radioactivity as described in the subscript to Table V, DNA concentrations were measured by extraction with 0.5 N perchloric acid, and measurement of the extinction at 260 nm. Calf thymus DNA served as standard. Each value stated is the average of at least two different experiments, and the experimental variation was ± 4% from the mean.

to increase with concentration of intercalating agent, eventually attaining a maximum constant value characteristic for each agent (Figure 3). From this, a value for the maximum number of primary binding sites per nucleotide can be estimated, and the good agreement with values determined by spectroscopic methods (Table VI), where available (39,40,41), indicates that stimulation arises primarily via intercalation of the agents into DNA.

Moreover, no stimulation is observed with structural analogues of intercalating agents bearing bulky substituents (Figure 4), which prevent intercalation on steric grounds (42,43, 44) or with oligocations lacking a planar aromatic ring system. The data summarized in Table VII indicate that neither intercalating agents, nor the types of cations mentioned above, have any stimulatory effect on methylation of free histone incubated

140

INTERCALATING AGENTS AND NON-INTERCALATING
STRUCTURAL ANALOGUES

INTERCALATING
(STIMULATE METHYLATION)

NON-INTERCALATING
(NO STIMULATION)

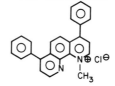

4,7 – dimethyl NMP

4,7 – diphenyl NMP

MIRACIL D

N' – methyl MIRACIL D

PROFLAVINE

Di – t – BUTYLPROFLAVINE

Fig. 4.

TABLE VII

Effect of Polar Aromatic Compounds on Histone Methylation with SAM in the Presence of Purified Histone Lysine Methyltransferase

Intercalating	Compound*	Inhibition of Methylation** (% of Control)
	Acridine Orange	57
	Actinomycin D	22 (15)
	Dinitroaniline reporter	0
	Ethidium Bromide	28 (22)
	Miracil D	12
	Naphthalimide reporter	9 (0)
	Naphthalene bisimide reporter	41
	2, 9-dimethyl NMP	10
	4, 7-dimethyl NMP	7
	2, 5, 6, 9-tetramethyl NMP	10
	Proflavine	28 (21)
Non-intercalating		
	4, 7-diphenyl NMP	85

* Concentration = 1.0 mM except for actinomycin D (0.16 mM)
** Numbers in parentheses are values obtained at one tenth the concentration used to obtain those not in parentheses.

Histone F_3(0.5 mg) was incubated with SAM-methyl-^{3}H (2.5 µ Ci), mercaptoethanol (0.05M) and lysine methyltransferase preparation (13) (300 µg protein) in 0.05 Tris buffer pH 9 for 10 minutes at 37°, total vol. = 0.5 ml. Various agents were added as neutral chloride salts except where stated. The reaction was stopped by addition of 2.5 ml 18% TCA, and after washing the precipitate with the same solution, histone was extracted with water and assayed for radioactivity as above. Values stated are the average of duplicates, and varied by ± 4% from the mean between repeated experiments.

with the purified histone lysine methyltransferase, and that many intercalating agents are even inhibitory under these conditions.

Analogues of the N-methylphenanthrolinium (NMP) cation provide a series of intercalating agents of similar structure which stimulate to different degrees. As is shown in Figure 5, the maximal values of stimulation of the 2, 9 and 4, 7-dimethyl

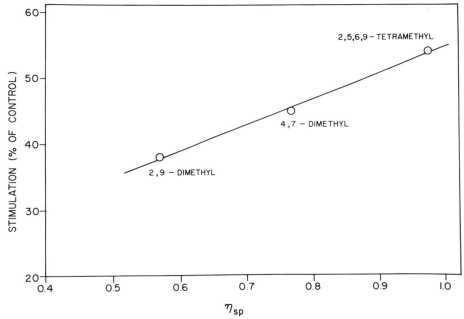

Fig. 5. Variation of maximal stimulation of histone methylation in nuclei (quoted from Table VI) with specific viscosity (values quoted for ref. 42) for methyl derivatives of the NMP cation.

and 2, 5, 6, 9-tetramethyl NMP cations correlate linearly with the maximal relative viscosity of their complexes with DNA (42). These correlations suggest that stimulation of histone methylation is primarily caused by the lengthening and distortion of DNA due to intercalation of cations between base pairs, and not to e.g. differential permeability of the nuclear membrane for intercalating agents.

A further observation which can be made is that while for most intercalating agents there are many fewer binding sites in chromatin than in free DNA (45), no difference can be found with the dinitroaniline reporter (40) and 4, 7-dimethyl NMP (comparing our data for chromatin in Table VI with those of Gabbay (42) for DNA). The observation that the chromosomal proteins apparently do not interfere with binding of the dinitroaniline reporter was first made by Simpson (40), and led him to suggest that the reporter molecule binds to the minor, the chromosomal proteins to the major, groove. However, this may be an over-simplification, since this reporter was subsequently shown to intercalate between the DNA base-pairs, rather than to actually bind in the minor groove (46). Having only a small aromatic ring system compared to the other intercalating agents, it would not be expected to occlude either groove of DNA to a serious extent. In comparison, intercalation of ethidium bromide places a bulky phenyl group in the major groove of DNA (47) and is indeed excluded from a large proportion of its DNA binding sites in chromatin by proteins binding wholly or partially to that groove (41). A further paradox occurs with actinomycin D, for which there is strong evidence of binding solely to the minor groove (48), and yet has a reduced number of binding sites in chromatin, compared to free DNA (39). The concept of intercalating agents being indicators for protein binding either to the major or minor groove in chromatin appears therefore confusing, and the chromosomal proteins may not in fact be restricted to the major groove of DNA in chromatin.

Polylysines.

Polylysine has been widely used as a probe of DNA structure in chromatin (49-54). Whereas it can bind to naked DNA to the extent where the lysine:DNA phosphate ratio is 1.0 (52), the saturation value of that ratio in polylysine-chromatin complexes appears to be approximately 0.5 (49). The thermal stability of DNA:polylysine complexes is markedly dependent on the length of the lysine sequence (53), but possible perturbations in DNA secondary structure in the complexes have been disputed.

Although in the presence of lysine oligomers with up to five lysine residues, or the oligocations spermine, spermidine or MethylGAG, no effect was observed on histone methylation in nuclei, lysine sequences of nine or more residues caused marked stimulation of methylation. As can be seen in Figure 6, maximal stimulation occurred at a lysine:DNA phosphate ratio of 0.5 for polylysines of various sequence length. The latter ratio is in agreement with the value reported previously for maximal binding of polylysine to chromatin (49). At higher ratios the degree of stimulation decreases rapidly and polylysines of large sequence length even inhibit methylation, probably by displacing histones from chromatin (27).

In contrast, the methylation of free histone by the purified methyltransferase was found to be inhibited, rather than stimulated, by polylysines of longer sequence than eight residues. With oligocations no effect was observed at any cation: DNA-phosphate ratio.

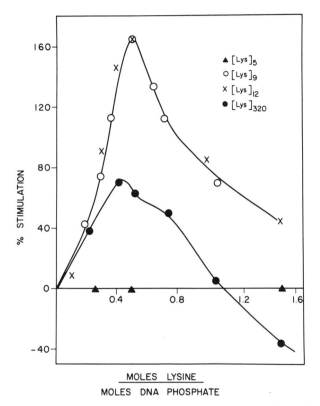

Fig. 6.   Stimulation of histone lysine methylation in nuclei in the presence of lysine polymers.   Histone methylation was assayed as described in the subscript to Table V.   The polycations were added as neutral bromide or acetate ([Lys]$_5$) salts.

TABLE VIII

Effect of Oligo- and Polylysines on the Stimulation of Histone Methylation in Nuclei by Intercal-ating Agents

Intercalating Agent	Concn. (mM)	% Stimulation			
		no polycation	(Lys)5*	(Lys)9*	(Lys)320**
None	--	--	0	117	63
Actinomycin D	0.1	56	36	168	107
Proflavine	0.5	72	72	151	109

* Concentration = 1 mM lysine
** Concentration = 1.5 mM lysine

Histone methylation in nuclei was assayed as described in the subscript to Table V. The values reported varied by ±4% from the mean between different experiments.

Competition studies with polylysines and intercalating agents (Table VIII) revealed that pentalysine antagonizes the stimulation of histone methylation by actinomycin D but does not affect that by proflavine, an intercalating agent which presumably occupies the major groove of DNA. These findings confirm that the former binds only to the minor groove of DNA in chromatin (55). The reverse effect occurs with polylysines of higher degrees of polymerization. Whereas the stimulation by polylysines of $n \geq 9$ and that by actinomycin D are approximately additive if combined, a considerable deviation from additivity is observed with a combination of these polycations and proflavine. This observation suggests that polylysines with nine residues or more bind principally in the major groove.

Our experiments suggest that the mode of binding to chromatin of oligolysines is markedly different from that of polylysine with 9 or more lysine residues. In agreement with this, it has been reported that biphasic melting curves for DNA/polylysine complexes are only observed with polylysines of $n \geq 8$ (53).

Since our evidence indicates that the stimulation of histone methylation in nuclei by intercalating agents is caused by the perturbation of the DNA structure which results from intercalation, it would seem reasonable to speculate that the stimulatory action observed with polylysines might also be attributable to distortion of DNA structure upon polylysine binding.

The fact that relatively small changes in DNA structure cause a rather conspicuous stimulation of histone methylation suggests that DNA, histones and methyltransferase must be situated in close proximity in premitotic chromatin, the enzyme presumably being present only at the end of S-phase and G_2 (56). Why such changes cause stimulation rather than inhibition is a complete mystery. Certain structural conditions will apparently either activate or inhibit the enzyme. Especially intriguing is the possibility that such changes may trigger previously nonexistent methylation in euchromatin, a question presently under investigation in this laboratory.

CONCLUSION

These studies have shown that the polar portions of the methylated histones are bound more tenaciously to DNA than those of other histones. It seems reasonable to assume that ε, N-methylation of lysine residues is at least a contributing factor. How important this binding is for the repression of undesired information, is not known, but it is clear that a variety of substances with widespread biological actions affect the histone methylation process. Most of these have been shown to interact with DNA in one or another way, and it is probably safe to assume that at this structural level a number of epigenetically important processes take place and that histone methylation is just one of these.

Acknowledgements

We wish to thank Drs. K. N. Scott and C. E. Westerman, Department of Radiology, University of Florida, for help in obtaining CMR spectra; Dr. E. J. Gabbay, Department of Chemistry, University of Florida, for samples of the Dinitroaniline, Naphthalimide and Naphthalene Bisimide reporter molecules; Professor W. S. Brey, Department of Chemistry, University of Florida, for helpful discussions and Mrs. M. C. Chow for expert technical assistance.

This work was supported by NIH grants CA 13408 and G 1419074.

References

1.   K. Murray. Biochemistry 3:10 (1964).

2.   W. K. Paik and S. Kim. Science 174:114 (1971).

3.   W. K. Paik and S. Kim. Biochem. Biophys. Res. Commun. 27: 479 (1967).

4.   V. K. Hempel, H. W. Lange, L. Birkhofer. Zeitschr. Naturforsch. Bl: 37 (1968).

5.   W. K. Paik and S. Kim. Biochem. Biophys. Res. Commun. 29: 14 (1967).

6.   E. L. Gershey, G. W. Haslett, G. Vidali, V. G. Allfrey. J. Biol. Chem. 244:4871 (1969).

7.   R. DeLange and E. L. Smith. Account Chem. Res. 5:368 (1972).

8.   P. Byvoet. Biochim. Biophys. Acta 238:375a (1971).

9.   G. R. Shepherd, J. M. Hardin and B. J. Noland. Arch. Biochem. Biophys. 143:1 (1971).

10.   P. Byvoet, G. R. Shepherd, J. M. Hardin and B. J. Noland. Arch. Biochem. Biophys. 148:558 (1972).

11. W. K. Paik and S. Kim. J. Biol. Chem. 243:2108 (1968).

12. W. K. Paik and S. Kim. J. Biol. Bhem. 245:1806 (1970).

13. W. K. Paik and S. Kim. J. Biol. Chem. 245:6010 (1970).

14. T. Tidwell, V. G. Allfrey and A. E. Mirsky. J. Biol. Chem. 243:707 (1968).

15. P. Byvoet. Arch. Biochem. Biophys. 148:558 (1972).

16. P. Byvoet. J. Mol. Biol. 17:311 (1966).

17. M. Noll. Nature 251:249 (1974).

18. A. L. Olins and D. E. Olins. Science 183:330 (1974).

19. C. Tanford. J. Amer. Chem. Soc. 84:4240 (1962).

20. E. M. Arnett, F. M. Jones III, M. Taagerpera, W. G. Henderson, J. L. Beauchamp, D. Holtz and R. W. Taft. J. Amer. Chem. Soc. 94:4724 (1972).

21. D. H. Aue, H. M. Webb and M. T. Bowers. J. Amer. Chem. Soc. 94:4726 (1972).

22. G. C. Levy and G. L. Nelson. In: "Carbon-13 Nuclear Magnetic Resonance for Organic Chemists", Wiley-Interscience, New York, Chapter 4 (1972).

23. K. Murray. J. Mol. Biol. 15:409 (1966).

24. H. H. Ohlenbusch, B. H. Oliver, D. Tuan and A. N. Tsvetikov J. Mol. Biol. 25:209 (1967).

25. K. Evans, P. Konigsberg, and D. R. Cole. Arch. Biochem. Biophys. 141:389 (1970).

26. J. E. Smart and J. Bonner. J. Mol. Biol. 58:651 (1971).

27. P. Byvoet, Submitted for publication.

28. E. W. Johns and S. Forrester. Eur. J. Biochem. 8:547 (1969).

29. P. Bohler, S. Stein, W. Dairman and S. Udenfriend. Arch. Biochem. Biophys. 155:213 (1973).

30. D. R. Van der Westhuyzen and C. Von Holt. FEBS Lett. 14: 333 (1971).

31. R. T. Simpson. Biochemistry 11:2003 (1972).

32. E. M. Bradbury and C. Crane-Robinson. In: "Histones and Nucleohistones", Plenum Press, D. M. P. Phillips, ed. (1971).

33. C. S. Baxter and P. Byvoet. Cancer Res. 34:1424 (1974).

34. C. S. Baxter and P. Byvoet. Cancer Res. 34:1418 (1974).

35. M. Friedman, K. H. Shull and E. Farber. Biochem. Biophys. Res. Commun. 34:857 (1969).

36. J. Chiu, C. Craddock, S. Getz and L. S. Hnilica. FEBS Lett. 33:247 (1973).

37. W. J. Pigram, W. Fuller and M. E. Davies. J. Mol. Biol. 80:361 (1973).

38. L. S. Lerman. Proc. Nat. Acad. Sci. USA 49:94 (1963).

39. L. Kleiman and R. C. Huang. J. Mol. Biol. 55:503 (1971).

40. R. T. Simpson. Biochemistry 9:4814 (1970).

41. P. F. Lurquin and V. G. Seligy. Biochem. Biophys. Res. Commun. 46:1399 (1972).

42. E. J. Gabbay, R. E. Scofield and C. S. Baxter. J. Amer. Chem. Soc. 95:7850 (1973).

43. I. B. Weinstein and E. Hirschberg. Prog. Mol. Subcell. Biol. 2:232 (1971).

44. W. Muller, D. M. Crothers and M. R. Waring. Eur. J. Biochem. 39:223 (1973).

45. P. Lurquin. Chem-Biol. Inter. 8:303 (1974).

46. F. Passero, E. J. Gabbay, B. Gaffney and T. Kurucsev. Macromolecules 3:158 (1970).

47. W. Fuller and M. J. Waring. Ber. Bunsenges. Phys. Chem. 68:805 (1964).

48. H. M. Sobell, S. C. Jain, T. D. Sakore and C. E. Nordman. Nature New Biol. 231:200 (1971).

49. R. J. Clark and G. Felsenfeld. Biochemistry 13:3622 (1974)

50. R. F. Itzhaki. Eur. J. Biochem. 47:27 (1974).

51. J. J. Li, C. Chang, M. Weiskopf, B. Brand and A. Rotter. Biopolymers 13:64 (1974).

52. D. E. Olins, A. L. Olins and P. H. von Hippel. J. Mol. Biol. 24:157 (1967).

53. D. E. Olins, A. L. Olins and P. H. von Hippel. J. Mol. Biol. 33:265 (1968).

54. M. Haynes, R. A. Garrett and W. B. Gratzer. Biochemistry 9:4410 (1970).

55. D. Carroll and M. R. Botchan. Biochem. Biophys. Res. Commun. 46:1681 (1972).

56. H. W. Lee, W. K. Paik and T. W. Borun. J. Biol. Chem. 248:4194 (1973).

MULTIPLE BINDING SITES FOR PROGESTERONE IN THE HEN OVIDUCT NUCLEUS: EVIDENCE THAT ACIDIC PROTEINS REPRESENT THE ACCEPTORS

Thomas C. Spelsberg, Robert Webster, and George M. Pikler
Department of Molecular Medicine
Mayo Clinic
Rochester, Minnesota 55901

Abstract

Steroid hormones, including progesterone (P), are known to bind with high affinity ($K_d \sim 10^{-10}$) to receptor proteins once they enter target cells. This complex (the P-receptor) then undergoes a temperature (or salt) dependent modification which allows it to migrate to the cell nucleus and to bind with a lesser affinity ($K_d \sim 10^{-8}$ to $\sim 10^{-9}$ M). The studies to be presented demonstrate that the in vitro nuclear binding of the P-receptor to isolated nuclei from the oviducts of laying hens has the same properties and requirements as was reported earlier for the immature chick system. Furthermore, using the partially purified P-receptor revealed that the differences in extent of binding to the nuclear material between a target tissue (oviduct) and other tissues (liver, spleen, or erythrocyte) were markedly dependent on the ionic conditions. The assay of the nuclear binding, using multiple levels of P-receptor, revealed the presence of more than one binding site in the oviduct nuclei. The hormone binding to each of the sites displayed a differential stability to increasing levels of ionic strength. Only the highest affinity binding site was capable of binding P-receptor under the higher ionic conditions. The multiple binding sites for the P-receptor were not found in the nuclei of a nontarget organ (spleen). Oviduct chromatin displayed the multiple binding sites as did whole nuclei. New techniques for assaying the binding of P-receptor to soluble nucleoproteins were developed. Using antibiotics to precipitate the DNA and nucleoproteins or the attaching of the latter to insoluble resins allowed the rapid assay of P-receptor binding with a minimal background. Using these techniques, binding to dehistonized oviduct chromatin (containing acidic protein fractions AP_3 and AP_4) demonstrated that the highest affinity binding site was still present. When the acidic protein AP_3 was removed from the DNA, this highest affinity binding site was lost. Analysis of the amounts of receptor and DNA per ovi-

duct cell suggests that only the highest affinity site is involved in the nuclear binding of progesterone. These results
concerning the AP_3 fraction corroborate earlier reports on progesterone binding in the chick oviduct.

There are no known biological functions assigned without
some doubt to the acidic (nonhistone) chromatin proteins, ex-
cluding enzyme activity (1). There has been speculation and
some evidence that these proteins play a role in the regulation
of gene activity via regulating the modulation of the rate
of transcription as well as in regulating the tissue specific
restrictions of the DNA in chromatin. The involvement of these
proteins in the maintenance of structure of the interphase
chromatin fiber and the metaphase chromosome has also been
suggested. In the past few years, another biological function
of the acidic chromatin proteins has been suggested: They may
serve as recognition sites for certain cytonucleoproteins which
appear to be involved in the transport of steroid hormones from
the cytoplasm to the nucleus of target cells (2-6). As soon as
the steroid hormone enters the cytoplasm of a target cell, it
forms a stereo-specific complex with high affinity with a "receptor" protein. This complex then migrates into the nucleus,
binds to chromatin, and causes alterations in the transcription
of the DNA. Part of this alteration of transcription appears
to be due to differential alteration in the restriction of the
DNA template. The outline of such a mechanism is shown in Figure 1. Within minutes after injection of a labeled steroid in-

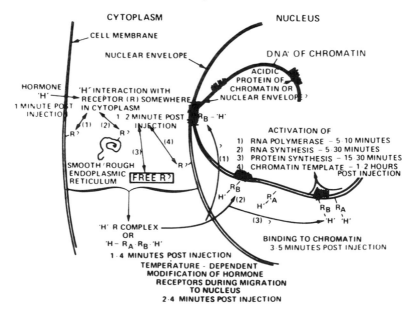

to an animal, the appearance of labeled steroid complexed to cytosol proteins and to the nuclear chromatin is observed.

In the past eight years, laboratory developments have allowed the formation of this steroid-receptor protein complex to be achieved in vitro. Also, the in vitro binding of this complex to isolated nuclei or chromatin has been achieved, which appears to require many of the same conditions as required in vivo. The in vitro nuclear bound steroid appears to have all of the same properties with respect to affinity, ionic dissociation of the steroid as a complex with its receptor, etc., as the complex which is formed in vivo or in whole cells.

Past work in this and in other laboratories using ^{3}H-progesterone complexed with the chick oviduct cytosol receptor and the nuclear components from various organs of the chick has suggested that 90% of the total nuclear binding in the in vitro system is associated with the nuclear chromatin (2-4,9). Under physiological ionic strength, the extent of binding of the progesterone-receptor complex appeared to be enhanced with the chromatin from the chick oviduct compared with that from spleen, heart, and other tissues. By selectively removing proteins from the chromatin as well as performing reconstitution of the chromosomal proteins to the DNA to form "hybrid" chromatin, evidence was obtained suggesting that the acidic (nonhistone) chromatin proteins represent the acceptor sites to which the progesterone-receptor complex binds. One fraction of acidic chromatin proteins, fraction AP_3, was found to contain this acceptor (3,4). The DNA-free proteins were unable to bind the progesterone-receptor complex. However, the proteins could be reannealed to the DNA with the subsequent renaturing of the binding capacity. Although the binding of the progesterone-receptor complex to pure DNA was observed, the binding to the DNA which was reannealed with the AP_3 protein fraction caused a 10 to 50 fold enhanced level of binding per unit mass of DNA.

Studies presented here represent a continuation of these earlier studies with the objective to identify and characterize the acceptor for the progesterone-receptor complex in oviduct nuclei of mature laying hens. The hen oviduct was substituted for the chick oviduct in these more recent studies in order to

Fig. 1. Model of the mechanism of action of steroid hormones in target cells. The R_A, R_B represent two species of receptors for progesterone reported by Schrader and O'Malley (13). The R_A is postulated to bind to the DNA while the R_B to the acidic protein of chromatin (taken from Spelsberg, ref. 4). (Reprinted with permission of Academic Press).

obtain sufficient masses of tissue for the ultimate purification of the acceptor.

MATERIALS AND METHODS

Isolation and Analyses of the Nuclear Materials

Various organs of adult laying hens were obtained from a
local produce company (Jones Produce, Rochester, Minnesota).
Several   minutes after the hens were bled to death, the various organs were obtained from the production line, stripped of
fat and connective tissue, sectioned into small pieces, and
frozen immediately in dry ice. These tissues were then stored
at -80°C in a Revco freezer until use. All subsequent steps
were performed at 4°C. The isolation and purification of organ
nuclei as well as chromatin were performed essentially as described elsewhere (2,10). Mature erythrocytes from these laying
hens were isolated as reported previously (11), and the nuclei
and chromatin isolated as the organ material. The various
fractions of chromatin, the acidic protein fractions AP_1 and
AP_2, histones, and nucleoacidic proteins (NAP) were isolated as
described elsewhere (3) (the procedure is outlined in Figure
16). Pure DNA was obtained by the method described by Spelsberg et al. (10). All preparations were monitored by chemical
analysis and thermal denaturation profiles as described previously (3,10).

Steroids

$(1,2-^3H)$ progesterone, 47.8 curies/mmole, was obtained
from New England Nuclear. The degree of chemical purity of
this material was checked by chromatography on thin layer
plates with benzene ethylacetate (70:30 v/v). The stock steroid solution in benzene was frozen and lyophilized to dryness
and then redissolved in the original volume with absolute alcohol and diluted 1:5 with distilled water. This solution was
then added directly to the cytosol preparations to obtain the
progesterone-receptor complexes as described in the next section.

Preparation of Progesterone-Receptor Complex

The oviducts from chicks treated for 20-25 days with di-
ethylstilbestrol (DES) were excised, stripped of fat and immediately homogenized in a Waring blender in three volumes (w/v)
of cold TESH buffer (0.01 M Tris HCl, 0.001 M EDTA and 0.012 M
thioglycerol, pH 7.4). Homogenization was carried out for 30
sec with a rheostat setting of 6. The homogenate was then
rehomogenized briefly in a Teflon pestle-glass homogenizer and

the homogenate then passed through two layers of cheese cloth. The solution was centrifuged at 20,000 x g_{av} for 20 min in a J-21 Beckman centrifuge, and the supernatant recentrifuged at 100,000 x g_{av} for 1 hour in a Beckman L3-50 ultracentrifuge. The 100,000 x g_{av} supernatant (cytosol) was then assessed for protein concentration by absorption at 280 mμ and by the Lowry method (12). The solution was then diluted to give a 20 mg protein/ml solution. To this cytosol was added 5 μl of the progesterone solution described in the above section, containing 80% water and 20% alcohol. This gives a concentration of about 2.09 x 10^{-8}M progesterone, which is about 1 μCi of radioactivity per ml. This labeled cytosol was then incubated for 2 hrs at 4°C and represents the crude (^{3}H)P-receptor. Usually these steps were followed by the addition of saturated ammonium sulfate in TESH buffer solution to a final concentration of 35% saturation with respect to ammonium sulfate. After 60 min, the material was collected by centrifugation at 20,000 x g_{av} for 10 min in a Sorvall HB-4 swinging bucket rotor. This procedure precipitates literally all of the hormone bound receptor, leaving the unbound hormone and the majority (95%) of the cytosol proteins in the supernatant (13). The supernatant was then removed and the walls washed gently with water. These pellets were either frozen at -80°C until needed or used immediately, whereupon they were resuspended in TESH buffer at one half the volume of the original cytosol solution, dialyzed in washed dialysis tubing against 20 volumes of TESH buffer at 4°C for 2 hrs, and then recentrifuged 10,000 x g_{av} for 5 min to sediment any insoluble material. This ammonium sulfate purified hormone-receptor complex was used directly in the nuclear binding assays. The integrity of the hormone-receptor complex was constantly monitored by sucrose gradient centrifugation and by the charcoal-dextran-bovine plasma albumin methods as described previously (13). Greater than 90% of the radioactivity was bound to a 4S sedimenting protein.

Assays for Nuclear Binding

The Standard Method: A standard or routine method for performing binding of the partially purified progesterone-receptor complex to nuclei or chromatin simply involved the addition of the labeled hormone-receptor complex to the nuclear material, incubating, washing, collecting on filters, and counting. Briefly, 1 ml reactions contained 25 μg of DNA as nuclei or chromatin, 10% glycerol in the case of nuclei only, one half the concentration of TESH buffer (or 0.005 M Tris HCl, 0.005 M EDTA, 0.006 M thioglycerol, pH 7.4), the necessary KCl, and water to bring it up to final volume. Nuclei isolated for hormone binding assays were resuspended the final time in

50% glycerol containing 1/5th TKM buffer (0.01 M Tris HCl, 0.005 M KCl, 0.004 M magnesium chloride). A 50% glycerol solution in the TKM solvent alone was used to bring the final concentration of glycerol to 10% in the reactions when using nuclei. The TESH buffer was added in varying amounts depending on the amount of receptor being added to keep the final concentration one half the strength of the original TESH buffer. The reactions were begun with the addition of the receptor; the reactions were then immediately vortexed and set in an ice bucket for 90 min. The reaction vessels were then centrifuged 600 x g_{av} for 5 min in a clinical centrifuge. The nuclear pellets were resuspended in 2 ml of a dilute Tris-EDTA buffer (0.002 M Tris HCl, 0.0001 M EDTA, pH 7.5). The nuclear material was resedimented and washed again similarly in 2 ml of the Tris-EDTA solution. Upon the third resuspension in 2 ml of Tris-EDTA solution, the nuclear material was collected on Millipore filters (0.45 micron pore size, 24 mm diameter, from the Millipore Corp., Bedford, Mass.). The reaction vessels and filters were washed with five more mls of the Tris-EDTA solution, the filters dried at room temperature and counted in a scintillation spectrometer in 5 ml of the toluene base PPO-POPOP solution. After counting, the filters were removed, dried thoroughly, and the DNA hydrolyzed by incubating the filters in 0.3 N HClO$_4$ for 30 min at 90°C. The cooled solutions were then analyzed for DNA by the diphenylamine reaction (14). The counts per minute per mg DNA (cpm/mg DNA) were then calculated. In some instances, the specific activity of the labeled hormone was used to calculate either the pmol labeled ^{3}H-P bound or free per mg DNA in each of the reaction solutions, or the molarities of the labeled progesterone (bound or free) were calculated.

Streptomycin Method: This method was developed in order to assay the binding of the progesterone-receptor complex to soluble nuclear fractions and will be described in detail elsewhere (15). This technique and the following one (cellulose method) avoided the long periods of centrifugation in the ultracentrifuge. The reactions were set up essentially as described above for chromatin. When assaying the binding to either chromatin, nucleoacidic protein (NAP) or DNA, no glycerol was included. In this method, however, 50 μg of DNA is required for optimal precipitation and recovery of the DNA. At the end of the 90 min incubation at 4°C, approximately 1 mg of streptomycin sulfate was added to each reaction followed by a slight mixing with the Vortex mixer. After sitting for 20-30 min at 4°C, the solutions were then centrifuged for 10 min at 2,000 x g_{av}, the supernatants removed, and the pellets washed twice with the same buffers described for the standard method except they contained 0.02% streptomycin. The nuclear material was collected on Millipore filters and counted as described above. Using highly purified streptomycin sulfate, less than 10-15% of the progesterone-receptor complex is dissociated under the concentrations used. About 60 to 80% of the DNA is recoverable in the assays.

Cellulose Method: As an alternate procedure to the streptomycin method to assay for hormone binding to soluble nuclear protein fractions, the attachment of these soluble nuclear fractions to cellulose was performed. Using modified conditions of those described by Litman (16), which are to be described in detail elsewhere (17), resins containing high levels of tightly bound DNA, chromatin or nuclear protein fractions are prepared. Resins containing 10-20 mg of DNA/g of resin can be achieved. However, in most of the experiments described here, the resins contained 2-10 mg of DNA/g of resin. Nuclei, chromatin, NAP, DNA and other partially deproteinized chromatin fractions bound to cellulose using ultraviolet light in the presence of absolute alcohol were subsequently washed in dilute buffers, lyophilized, and the dry resin weighed and assayed for DNA by the method of Burton (14). For each experiment, aliquots of the resin were weighed out which would be sufficient in DNA content to provide for a planned experiment. These resins were resuspended in the Tris-EDTA buffer and allowed to hydrate for several hours at 4°C before use. While mixing, aliquots of this hydrated resin were placed in reaction vessels to give a designated amount of DNA (which usually ranged from 20-25 μg DNA/vessel). The remaining solution and the binding assay were essentially as described under the standard method. Similar amounts of pure cellulose (containing no DNA) as used with the resin containing DNA, were used as blanks in the hormone binding assays. Under the binding assay conditions, less than 10% of the DNA in any of the cellulose resins was lost. The use of these nuclear protein fractions, immobilized on cellulose, allowed rapid dissociation of the proteins from the DNA using high salt-urea or guanidine hydrochloride solvents followed by rapid assay of the capacity of these treated samples to bind the progesterone-receptor complex. These cellulose bound nuclear fractions in the lyophilized state were found to be stable for months at room temperature. Figure 2 shows scanning electron micrographs of some of these resins. Details of the preparations of these nuclear resins will be described elsewhere (17).

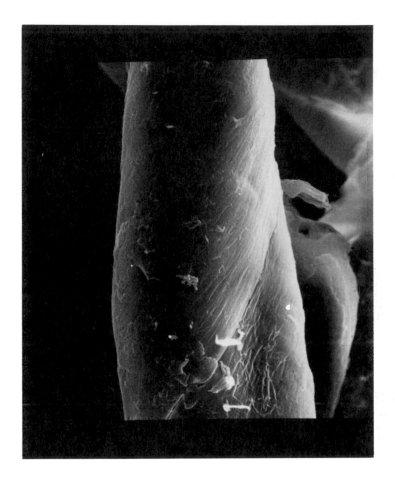

Fig. 2a. Scanning electronmicrographs of pure cellulose.
These micrographs were taken with an ETEC U-I Scanning electron
microscope. Samples of the resin were coated in a Hummer II
D. C. Sputterer with gold/palladium, 80/20 (w/w). Processing
and pictures were kindly performed by Dr. Harold Moses, Dept.
of Pathology and Anatomy, Mayo Clinic, Rochester, Minnesota.

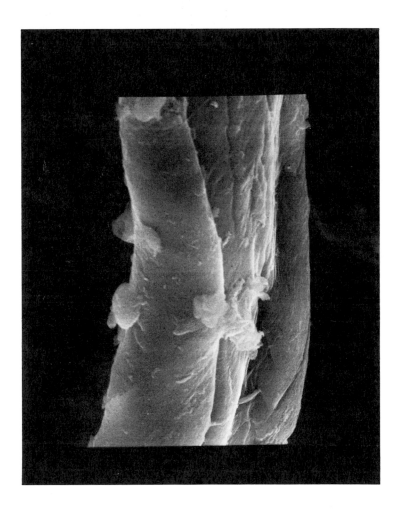

Fig. 2b. Scanning electronmicrographs of nuclei–cellulose.
These micrographs were taken with an ETEC U–I Scanning electron
microscope. Samples of the resin were coated in a Hummer II
D. C. Sputterer with gold/palladium, 80/20 (w/w). Processing
and pictures were kindly performed by Dr. Harold Moses, Dept.
of Pathology and Anatomy, Mayo Clinic, Rochester, Minnesota.

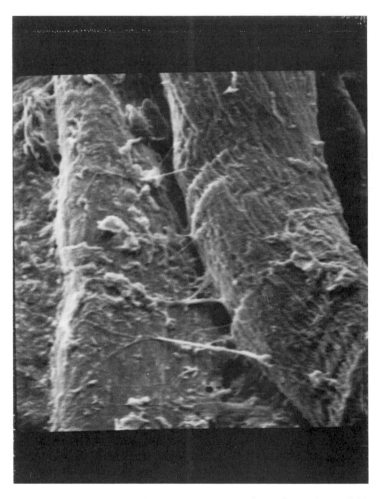

Fig. 2c.   Scanning electronmicrographs of nucleoacidic protein (NAP)-cellulose.

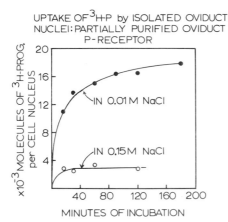

UPTAKE OF ^{3}H-P by ISOLATED OVIDUCT NUCLEI: PARTIALLY PURIFIED OVIDUCT P-RECEPTOR

MINUTES OF INCUBATION

Fig. 3. Kinetics of binding of the ^{3}H-progesterone-receptor complex to hen oviduct nuclei. Hen oviduct nuclei were isolated and purified as described in the Methods. After incubation at 4°C, the nuclei were pelleted in the clinical centrifuge, washed twice in a cold solution containing 2 mM Tris HCl plus 0.1 mM EDTA pH 7.5 (2 ml each wash). The nuclei were then resuspended a final time in the same solvent and transferred and collected on Millipore filters with two 1.0 ml washes with the Tris-EDTA buffer. The filters were then dried under a heat lamp and counted in a scintillation spectrometer. The range of three replicates of analyses are shown. The DNA per filter was assayed and the counts per mg DNA calculated as described in Methods. The specific activity of the ^{3}H-progesterone (44.8 Ci/mmole) and the 2.5×10^{-12} g DNA/cell were used to calculate the molecules of bound progesterone per cell nucleus. Assays were run with 25 µg DNA as nuclei with 200 µl of the partially purified receptor per assay using the standard method. The assays were run under (●) 0.01 M NaCl or (o) 0.15 M NaCl.

RESULTS

Properties and Requirements of the In Vitro Nuclear Binding

Many studies were performed to assess the requirements and nature of the in vitro binding of the progesterone-receptor complex to hen oviduct nuclei and chromatin. Figure 3 shows the kinetics of the nuclear uptake and binding of the ^{3}H-P-receptor complex. When assayed under very low ionic conditions the nuclear uptake of the ^{3}H-P-receptor complex began to equilibrate between 90 and 120 min of incubation. However, under the higher ionic conditions (0.15 M sodium chloride) the equilibrium of nuclear uptake was achieved within 30 min of incubation. Since all subsequent studies were to be performed between levels of ionic strength of 0.05 M and 0.15 M monovalent salt, all subsequent incubations were performed at 4°C for 90 min. It was found that the ^{3}H-P-receptor complex, incubated alone or with nuclei or chromatin for 90 min at 4°C, remained essentially intact as measured by sucrose gradient sedimentation analysis and charcoal binding assay (see Methods). To test the requirement of the receptor protein for nuclear binding, hen oviduct nuclei were incubated with ^{3}H-P either in buffer or complexed to its receptor protein. The amount of ^{3}H-P in buffer was determined by assessing the level of activity in the ^{3}H-P-receptor solution and adding the equivalent amount of radioactivity (as progesterone) to the buffer solution. As shown in Figure 4, the nuclear uptake and binding of the radioactive hormone was negligible when the receptor was omitted

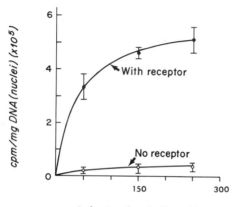

Fig. 4. Binding of ^{3}H-progesterone to hen oviduct nuclei using (•) ^{3}H-P complexed to its oviduct cytosol receptor or (o) ^{3}H-P in TESH buffer only. Incubations were carried out as described in Methods under 0.05 M NaCl using the standard method. About 25 µg DNA as nuclei per assay were used. The ranges of three replicates of analyses are shown.

from the reaction. Consequently, it appears that the nuclear uptake and binding of the labeled progesterone requires the receptor from the target tissue (the oviduct). Similar negligible uptake and binding of ^{3}H-P was observed when the hormone was incubated with the cytosol of a nontarget organ, e.g., spleen cytosol which contains no detectable receptor.

It is known for many in vitro steroid hormone-nuclear binding systems that the receptor needs to be "activated" either by mild heat treatment or by exposure to higher ionic strength before it is bound to isolated nuclei (7,18,19). It has also been reported that ammonium sulfate treatment of the progesterone receptor from the chick oviduct cytosol appears to simulate this activation process. As shown in Figure 5, the crude receptor preparation, not incubated at the higher temperature, shows a minimal uptake into the hen oviduct nuclei. The partially purified (by ammonium sulfate precipitation) progesterone receptor from the chick oviduct binds to nuclei under 4°C conditions as well as does the crude preparation (whole cytosol) which was incubated under 23°C conditions.

The increased binding of the partially purified receptor to the oviduct nuclei at 23°C as opposed to 4°C has previously been explained by Buller et al. (19) and has been shown experimentally in our laboratory to be caused by the opening of new binding sites in the nucleus. Nuclei incubated at 23°C for 90 min show proteolytic activity with respect to degradation of histone species, as well as a derepression of the chromatin DNA with respect to its capacity to serve as a template for DNA-dependent RNA synthesis in vitro (20). The use of the partially purified progesterone-receptor complex as opposed to the crude cytosol preparation in the binding assays was found to be more satisfactory for several reasons. Firstly, the stability of the hormone-receptor complex was much better in the former than in the latter. Also, the integrity of the nuclear chromatin was better preserved after incubations with the partly purified receptor preparation compared to incubation in the crude cytosol even under the 4°C conditions. Lastly, in assays using relatively low ionic conditions, the amount of cytosol protein complexing to the chromatin was found to be much less with the partially purified receptor preparation compared to the crude cytosol preparation. Under the higher ionic conditions of incubation (0.15 M KCl), the amount of protein complexed to the chromatin using either receptor preparation was minimized as reported previously (2). Consequently, all subsequent nuclear binding assays were performed with the partially purified receptor under the 4°C conditions.

Identification and Characterization of the Multiple Binding Sites in Oviduct Nuclei

The incorporation of numerous levels of the ^{3}H-P-receptor complex in the binding assay experiments was performed to obtain multiple points for analyses of the binding site affinity in the nucleus. As shown in Figure 6, these studies revealed the presence of several classes of binding sites. Since increasing levels of ionic strength cause the reduction in total binding of the ^{3}H-P-receptor complex to nuclei (see refs. 2, 7, 8 and 13 and Figure 3), similar experiments were performed under 0.10 M and 0.15 M KCl. As also shown in Figure 6, the increase in the ionic strength from 0.05 M to 0.10 M KCl during the binding assay caused a selective loss of binding of one class of sites. The increase of ionic strength from 0.10 to 0.15 M KCl caused the elimination of two more classes of binding sites. Further increases in ionic strength caused a rapid decrease in the level of radioactivity to background levels at 0.2 M KCl. It should be noted here that the TESH buffer in the reaction adds an equivalent of 0.02 M KCl as determined by conductivity. In any case, these results suggest that the various classes of binding sites in oviduct nuclei for the ^{3}H-P-

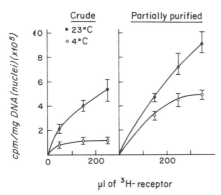

Fig. 5. Binding of ^{3}H-progesterone-receptor complex to hen oviduct nuclei using crude cytosol (100,000 x g_{av} supernatant of oviduct homogenates) or the partially purified receptor (35% $(NH_4)_2$ SO_4 precipitate). The assays were performed as described in Methods, using the standard method with 0.05 M NaCl and 25 µg DNA (as nuclei) per reaction. The temperature of each incubation was (•) 23°C or (o) 4°C. The ranges of three replicates of analysis are shown.

Fig. 6. Binding of ^{3}H-progesterone-receptor complex to hen oviduct nuclei: effects of ionic strength. The assays were performed using the standard method as described in Methods under (•) 0.05 M KCl, (o) 0.10 M KCl, and (x) 0.15 M KCl conditions. About 25 µg DNA per assay were used. The ranges of three replicates of analyses are shown.

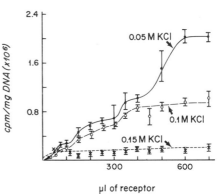

µl of receptor

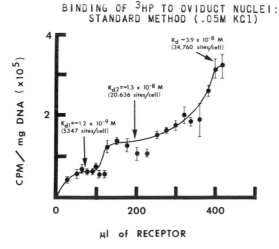

Fig. 7. (at left) Binding of ^{3}H-progesterone, complexed with its receptor, to hen oviduct nuclei using the standard method under 0.05 M KCl conditions as described in Methods. About 25 µg DNA per assay were used. The equilibrium dissociation constants and the number of binding sites per cell for the first two plateaus were calculated from the Scatchard plot (see

(continued on next page)

receptor complex are electrostatic in nature and that each can be selectively dissociated with specific levels of ionic strength.

Figure 7 shows the results of a similar type of experiment in which multiple levels of the hormone-receptor complex between 0 and 400 µl were assayed to identify better the class or classes of sites representing the highest affinity binding. In addition, this ratio of receptor to DNA better represents the in vivo conditions. The amount of recoverable receptor obtained per gram of tissue, as well as the amount of DNA per gram of tissue, indicates that the level of active receptor in whole cells represents about 30-50 µl per 25 µg of DNA (levels used in these experimetns). This level of receptor in our assays binds only the highest affinity class of acceptor. Scatchard plot analysis of the first two classes of binding sites suggests that the first class of binding sites has an equilibrium dissociation constant K_d = 1.2 x 10^{-9} M, representing 5.347 sites per cell nucleus (Figure 8). The second class

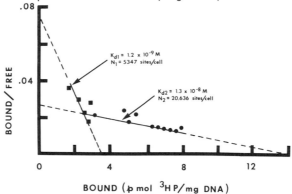

Fig. 8. Scatchard plot analysis of ^{3}H-progesterone-receptor binding to oviduct nuclei in 0.05 M KCl. Binding was carried out as described in Methods with 25 µg DNA per 1.0 ml assay mixture. The data were taken from Fig. 7. The number of binding sites was calculated assuming 2.5 pg DNA per oviduct nucleus. The K_d was calculated by extrapolating the amount of bound receptor to a one liter reaction volume, thus making the K_d = M/25 mg DNA.

Fig. 7. continued. Figure 8). The values for the third plateau were estimated from a hyperbolic plot of bound versus free ^{3}H-P receptor complex. The K_d was determined from the half saturation value from this plot. The DNA content per hen organ cell was taken as 2.5 x 10^{-12} g per cell. The ranges of three replicates of analyses are shown.

of binding sites has a 10-fold lower affinity with an equilibrium dissociation constant K_d = 1.3 x 10^{-8} M, representing
20,636 sites per cell nucleus. Scatchard plot analysis of the
third class of binding sites could not be achieved, so estimates of the equilibrium dissociation constant K_d, 3.9 x 10^{-8} M,
and the number of sites per cell, 34,760, were roughly calculated from a hyperbolic plot of bound versus free receptor from
these experiments using the half saturation point for the K_d
estimation.

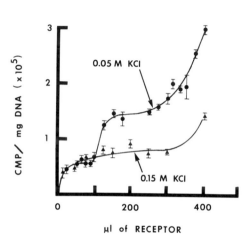

Fig. 9. The effects of
ionic strength on the binding of ^{3}H-progesterone-re-
ceptor complex to hen oviduct nuclei. These experiments were assayed essen-
tially as described in the
Methods and the legend of
Fig. 7 with the exception
that the assay conditions
contained either 0.05 M KCl
final concentration or 0.15
M KCl. All subsequent
treatments including washes
are essentially the same as
described in the legend of
Fig. 7. The ranges of 3
replicates of analyses are
given.

Figure 9 shows the effects of increasing the ionic conditions during the assay in the same range of receptor levels.
Clearly the 0.15 M KCl condition eliminates two classes of
binding sites, leaving only that class of binding sites representing the highest affinity binding. The possible inclusion
of a few sites from the second class of binding sites may occur.
Figure 10 shows a Scatchard plot analysis of the binding under
the higher ionic conditions to the oviduct nuclei. The equilibrium dissociation constant (K_d) of 6.24 x 10^{-9} M with the
number of sites of 10,090/cell nucleus is fairly close to that
observed for the highest affinity class of binding sites analyzed under low ionic conditions. A somewhat reduced level of
affinity of binding with a slight increase in the number of
sites per cell is measured. The increased values probably represent the inclusion of a few of the sites of the second highest affinity group. However, one cannot rule out the unmasking
of some of the sites on the nuclear chromatin by the higher
ionic strength, with the additional decrease in affinity of
the hormone-receptor complex to the highest affinity class of

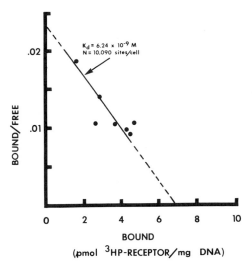

Fig. 10. Scatchard plot analysis of binding of the ^{3}H-progesterone-receptor complex to oviduct nuclei in 0.15 M KCl. The data were taken from Fig. 9. The binding and calculations are described in the legend of Fig. 8.

Fig. 11. Binding of ^{3}H-progesterone-receptor complex to the chromatin of hen oviduct and spleen using the streptomycin method under 0.10 M KCl conditions. The assay of this binding was carried out essentially as described in Methods and the legend of Fig. 3 with the exception that the streptomycin method was substituted for the standard method. Under this streptomycin method, 50-60 µg of DNA as hen oviduct or spleen nuclei were used for each assay. After the incubations, 1 mg of pure streptomycin as a 1% solution in water was added with slight stirring. The reactions were allowed to sit for 20 min at 4°C and then were sedimented in a clinical centrifuge. The pellets were washed twice as previously described in 2 ml of the Tris-EDTA buffer containing 0.02% streptomycin at 4°C. The pellets were then collected on Millipore filters and assayed as described in the Methods. The ranges of 2 replicates of analyses are shown.

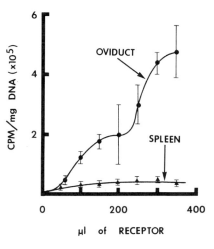

binding sites via the increased ionic conditions of the solvent.

Subsequent studies with isolated chromatin demonstrated that the multiple classes of binding sites were indeed of chromatin origin. Using the alternate streptomycin and the cellulose methods, the highest affinity class of binding sites as well as the weaker affinity binding sites were also observed in oviduct chromatin (Figures 11 and 12). Interestingly, these

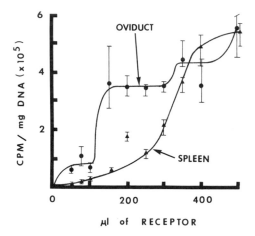

Fig. 12. The binding of ^{3}H-progesterone-receptor complex to the chromatin of hen oviduct and spleen using the cellulose method under 0.05 M KCl conditions. Hen oviduct and spleen chromatin were isolated and attached to purified cellulose as described in Methods. The chromatin-bound cellulose was then assayed for DNA content, and the mg of DNA/g of resin calculated. Before each experiment, excess resin was weighed out and suspended in a known volume of buffer containing 2 mM Tris HCl + 0.1 mM EDTA (pH 7.5) so that the concentration of DNA was known. The resin was allowed to hydrate in the buffer solution for several hours, using a slow stirring magnetic stirrer; then aliquots of the suspension were placed in individual reaction tubes to give the equivalent of 20-30 µg DNA per vessel. This was followed by the other reagents used in the binding assays, as described in the Methods section. During the 90 min reaction period at 4°C, the tubes were occasionally mixed with a Vortex stirrer (every 15 min). At the end of the assay the resin was sedimented in a clinical centrifuge, resuspended, washed, and collected on Millipore filters as described with the standard or streptomycin methods. After counting in a liquid scintillation spectrometer, the filters were removed from the fluor and the DNA per filter was assayed as described in the Methods. The ranges of 3 replicates of analyses of each assay are shown.

multiple classes of sites, especially those of the higher affinity, were not observed in the spleen chromatin as shown in Figures 11 and 12. Other tissues and organs of the hen, such as erythrocytes, also displayed no multiple classes of binding sites and no high affinity binding site.

More thorough analyses of the highest affinity binding site in oviduct nuclei or chromatin using the streptomycin or cellulose methods are summarized in Table I. Using the higher

TABLE I

Scatchard Plot Analysis of ^{3}H-Progesterone-Receptor Binding to Oviduct Nuclei and Chromatin in Using the Different Methods of Assay

Oviduct subcellular fraction	Method	KCl	K_d	N(sites/cell)
Nuclei	standard	0.05 M	1.2×10^{-9} M (First)	5,347
			1.3×10^{-8} M (Second)	20,636
Nuclei	standard	0.15 M	6.24×10^{-9} M	10,090
Chromatin	streptomycin	0.15 M	6.25×10^{-9} M	5,481
Nuclei	cellulose	0.15 M	1.91×10^{-8} M	15,300
Chromatin	cellulose	0.15 M	1.31×10^{-8} M	12,050

The assays for binding of ^{3}H-P-receptor complex to the nuclear fractions were performed with the three different methods which are described in the Methods and Materials.

ionic conditions (0.15 M KCl) to look selectively at the highest affinity class of binding sites, it was found that the
equilibrium dissociation constants as well as the number of
sites per cell under these alternate methods were roughly
equivalent to the highest affinity class of binding sites observed with the standard method with isolated oviduct nuclei.
The reduced affinity of binding as well as increased number of
binding sites per nucleus obtained with the cellulose method is
probably a result of the alcohol and ultraviolet light treatment and/or the lowered ionic conditions caused by the presence
of high levels of cellulose in each reaction. The latter was
detected using the conductivity measurements. In any case, the
alternate methods and conditions for analysis of nuclear binding by the ^{3}H-P-receptor complex detect the multiple classes of
binding sites as does the standard method. The highest affinity class of binding sites does survive the various conditions
and treatments described in this text. In addition, the multiple classes of binding sites observed with whole nuclei are
also observed and probably reside with the nuclear chromatin.

Subnuclear Localization of ^{3}H-P-Receptor Complex Binding in Oviduct Nuclei

There have been many reports of the fractionation of nuclear chromatin into active and inactive fractions. One of the
most popular techniques is one of shearing or sonicating isolated nuclei with the subsequent separation of the nuclear
fractions on sucrose gradients (21,22). Figures 13 and 14 show

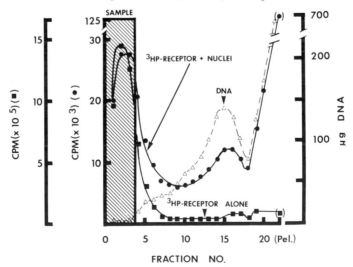

Fig. 13.   Caption on next page.

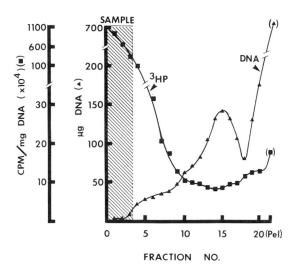

FRACTION NO.

Fig. 14. Distribution in a sucrose gradient of nuclear bound ^{3}H-progesterone-receptor complex and nuclear DNA from sheared oviduct nuclei. Isolated hen oviduct nuclei were incubated with the partially purified ^{3}H-progesterone receptor under 0.05 M KCl conditions using the standard method. They were then washed, sheared and assayed on the sucrose gradient as described in the legend of Fig. 13. The CPM per gm DNA was calculated and plotted.

Fig. 13. Distribution in sucrose gradients of ^{3}H-progesterone-receptor complex either incubated alone or with sheared nuclei for 90 min at 4°C under 0.05 M KCl using the standard method. Partially purified ^{3}H-progesterone-receptor complex was either layered directly on a sucrose gradient and centrifuged or incubated with oviduct nuclei, which were then washed, sheared, and centrifuged through a sucrose gradient. The nuclei were sheared in 0.01 M Tris, 0.01 mM MgCl$_2$, 5% glycerol (pH 8.0) with a Virtis homogenizer run at 15,000 rpm for 2 min. The homogenate was layered on a 0.1 M to 1.7 M sucrose gradient in 0.01 M Tris, 5% glycerol (pH 8.0). The gradients were centrifuged in the SW 27 rotor at 82,000 x g$_{av}$ for 2 hrs. The gradients were collected in 2.0 ml fractions. Aliquots were taken for the determination of DNA by the diphenylamine method of Burton (14) after precipitation with cold perchloric acid and for the determination of radioactivity in spectrofluor, containing Bio Solv solution.

results of typical experiments performed in this laboratory wherein isolated nuclei were complexed with the ^{3}H-P-receptor complex either before or after shear, and the subsequent nuclear material separated on a sucrose gradient. Fractions were collected, the DNA estimated by absorption at 260 mμ and by diphenylamine, and the fractions counted for presence of labeled progesterone. Figure 13 displays the raw data of such an experiment. It can be seen that much of the radioactivity remains in the sample with or without the presence of the nuclear material. The distribution of the level of radioactivity in the gradient followed the distribution of the chromatin DNA. Figure 14 shows the specific activity as the radioactivity per mg DNA of binding of the hormone to chromatin. The hormone appears to be fairly well distributed throughout the gradient with the exception of the upper portion. However, the upper fractions of the gradient represent primarily free steroid-receptor complex (as shown in Figure 13). Increased periods of centrifugation as well as inclusion of divalent cations, conditions which cause the sedimentation of most of the nuclear DNA to the pellet, failed to alter the pattern of radioactivity in the upper part of the gradient (fractions 1-8).

It appeared from these studies, which were performed under a variety of conditions, that there were no specific and localized subnuclear sites of binding of the hormone-receptor complex, at least as determined by these techniques. As a final check in some experiments, various fractions of the gradients were collected. The fraction tubes 5-11 were collected, pooled and labeled fraction I. Tubes 12-18 were collected, pooled and labeled fraction II; and tubes 19 and 20 plus the pellet (resuspended in 4 ml of Tris buffer) labeled fraction III. These fractions were then concentrated as described in the legend of Figure 15 and assayed for hormone binding capacity. The streptomycin method was used since fraction I was too soluble to be assayed by the standard method. It can be seen in Figure 15 that the binding capacities of the fractions were slightly but not markedly different. A series of experiments involving a variety of binding conditions using the various methods for assaying the hormone binding resulted in a similar observation as shown in Figure 15.

The Identification of the Nuclear Acceptor Site

Since the subnuclear localization of the binding sites for the ^{3}H-progesterone-receptor complex was not achieved, whole chromatin was required to be used as a starting point for the isolation and identification of the nuclear acceptor for the progesterone-receptor complex. The acceptor of interest here

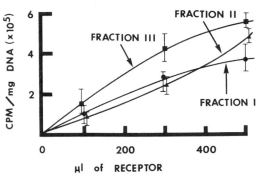

Fig. 15. Binding of ^{3}H-progesterone-receptor complex to hen oviduct nuclear subfractions using the streptomycin method under 0.15 M KCl conditions. The three nuclear subfractions were obtained from sucrose gradients (described in the legend of Fig. 13) by pooling tubes 5-11 for fraction I, tubes 12-18 for fraction II, and tubes 19 and 20 and the pellet for fraction III. The subfractions were concentrated by sedimentation in an ultracentrifuge and resuspended in small volumes of buffer (2 mM Tris HCl + 0.1 mM EDTA, pH 7.5). Fifty μg of DNA of each of these fractions were added to a series of tubes followed by increasing levels of the hormone-receptor complex. The assays were carried out essentially as that described in Methods. The ranges of 3 replicates of analysis of each assay are shown.

Figure 16. Scheme for Gross Fractionation of Chromosomal Proteins

Chromatin-Cellulose

2.0 M NaCl (pH 6.0)
2,000 x g 5 min
——————————→ supernatant
twice more   (histones)

Pellet
(dehistonized chromatin-cellulose)
(DNA-AP$_1$, AP$_2$, AP$_3$, AP$_4$-cellulose)

2.0 M NaCl + 5.0 urea (pH 6.0)
2,000 x g
——————————→ supernatant
                (AP$_1$ + AP$_2$)

Pellet
(DNA-AP$_3$, AP$_4$-cellulose)
0.01 M Tris HCl (pH 7.5) and dialysis (4 hr)

NAP-cellulose
(nucleoacidic proteins)
(DNA-AP$_3$-AP$_4$-cellulose)

was that involved in the highest affinity binding site detected
using either very low levels of receptor in the binding assays
or higher levels of receptor under higher ionic conditions
(0.15 M KCl). Initially, chromatin proteins were selectively
dissociated from the chromatin-cellulose resin according to the
scheme described in Figure 16, which is based on earlier reports (10,3). The selective removal of these groups of chromosomal proteins were monitored by gel electrophoresis (23,24).
The removal of the histones using 2.0 M NaCl (pH 6.0) was
essentially complete, with no detectable residual histone remaining. Subsequent removal of acidic protein fractions AP_1
and AP_2 were noted by the gel patterns of the extracts, as well
as of the residual material, and by quantitative analysis.
These analyzed preparations were then assayed for hormone binding capacity under the 0.15 M KCl conditions which remove all
but the highest affinity class of binding sites. As shown in
Figure 17, the removal of histones from chromatin failed to
alter the extent of binding of the ^{3}H-P-receptor complex compared to that of whole chromatin. However, the subsequent removal of acidic protein fractions AP_1 and AP_2 caused a marked
enhancement in extent of binding under these conditions. Since
the binding was carried out under relatively low receptor levels together with the higher ionic conditions, the major component of this binding is due to the highest affinity class of
binding sites as shown earlier. Repeated experiments using the
cellulose or streptomycin methods supported the results of
Figure 17 in that the removal of histones fails to unmask any
of the acceptor sites belonging to the high affinity class of
binding sites.

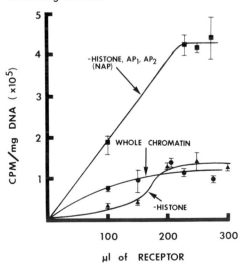

Fig. 17. Binding of ^{3}H-pro-
gesterone-receptor complex
to hen oviduct chromatin
preparations using the
cellulose method under 0.15
M KCl conditions. The as-
say of the binding of the
hormone-receptor complex
to the oviduct chromatin
preparation is essentially
that described in the legend
of Fig. 12 and Methods.
The chromatin preparations
were treated as described
in Fig. 16. Briefly, resin
was treated with 2 M NaCl
(pH 6.0) three times using

(continued on next page)

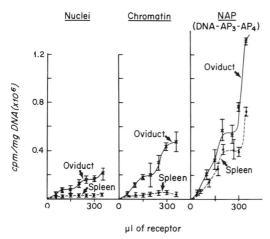

Fig. 18. Binding of the ^{3}H-progesterone-receptor complex to hen oviduct and spleen nuclear preparations using the streptomycin method under 0.10 M KCl conditions with 50 µg DNA per assay. The nuclei, chromatin and NAP were prepared as described in Methods. The ranges of 3 replicates of analyses for each assay are shown.

Other studies using the streptomycin method under lower ionic conditions also demonstrate that the nucleoacidic protein fraction contains a greater number of the higher affinity class of binding sites (Figure 18). Many of these sites are apparently masked in intact chromatin or nuclei. Interestingly, Figure 18 shows that the nuclei and chromatin of spleen do not display any of the high affinity classes of binding sites; however, the nucleoacidic proteins of spleen do reveal the presence of the high affinity class of binding sites almost equivalent to that of the oviduct. Pure DNA shows minimal binding under these conditions (about 1/10 that of the NAP).

Table II shows the results of similar types of studies with the objective to quantitate the extent of masking of the

---

Fig. 17 - continued below

20 volumes of buffered solvent per volume of packed cellulose resin. After each extraction, the resin was sedimented in a clinical centrifuge and resuspended again in each subsequent extraction. The resin was then washed several times in 2 mM Tris-0.1 mM EDTA (pH 7.5) solution. This treated chromatin resin preparation was designated as a dehistonized chromatin preparation with 90% of the original nonhistone proteins but no histone remaining. To obtain the NAP, which is chromatin deficient in histone, AP_1 and AP_2, a similar aliquot of whole chromatin was extracted in a solution containing 2 M NaCl + 5 M urea + 0.01 M phosphate buffer (pH 6.0). This extraction was essentially the same as that used to obtain the dehistonized chromatin resin. The NAP-cellulose showed no histones and only 50% of the total original nonhistone protein remaining. The ranges of 3 replicates of analysis of each assay are shown.

TABLE II

Masking of Acceptor Sites in Hen Organ Chromatin

Nuclear Fraction	At saturation of highest affinity binding site			% of sites exposed
	$\dfrac{cpm}{mg\ DNA}$	$\dfrac{pmoles\ bound}{mg\ DNA}$	$\dfrac{molecules\ bound}{cell}$	
1. Oviduct				
Chromatin	125,000	4.94	7,441	29%
Dehistonized Chromatin ($DNA + AP_1 + AP_2 + AP_3 + AP_4$)	156,000	6.16	9,278	37%
Nucleoacidic Protein ($DNA + AP_3 + AP_4$)	425,000	16.79	25,290	100%
2. Spleen				
Chromatin	25,000	0.99	<1,488	<6%
Nucleoacidic Protein	405,000	46.00	24,106	100%

Data were taken from Figures 17 and 18. The pmoles of ^{3}H-P bound/mg DNA were calculated using a specific activity of 47.8 Ci/mmole for the ^{3}H-progesterone. The molecules per cell were calculated based on a 2.5 x 10^{-12} g DNA/cell. The percents of sites exposed were calculated using the binding to the nucleoacidic protein as 100%.

acceptor sites for the ^{3}H-progesterone-receptor complex of the highest affinity class. In oviduct chromatin, only about 30 per cent of the total number of acceptor sites appear to be available for binding the progesterone-receptor complex. The removal of histones only increased by 6% the number of these sites, whereas removal of acidic protein fractions AP_1 and AP_2 resulted in the unmasking of 60% of the total high affinity binding sites. Interestingly, the nontarget tissue chromatin (spleen) demonstrated less than 6% of its total number of high affinity binding sites exposed in its intact chromatin compared to its nucleoacidic protein. This is preliminary evidence suggesting that nontarget tissues of a hormone may also contain the high affinity class of binding sites (the acceptors), but that these acceptors are completely masked. In addition, these acceptors appear to be associated with the acidic proteins of chromatin.

Another technique was established for the identification and isolation of the nuclear acceptor sites belonging to the high affinity binding group. High capacity oviduct chromatin-cellulose preparation was made and placed in a column. This resin was then subjected to a gradient of guanidine hydrochloride in a solvent containing a high concentration of a reducing agent at pH 8.5. Multiple peaks of protein eluted from this column as shown in Figure 19. The majority of the protein was

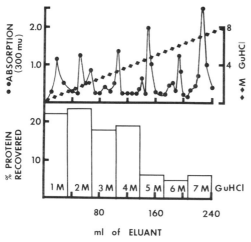

Fig. 19. Selective dissociation of proteins from chromatin-cellulose resins. Hen oviduct chromatin-cellulose resin was prepared as described in Methods. Twenty grams of this resin containing approximately 60 mg of DNA as chromatin was re-suspended in 100 ml of cold solution containing 0.1 M HSETOH + 0.05 M sodium bisulfite + 0.01 M Tris HCl (pH 8.5) and allowed to hydrate for 6 hrs with gentle stirring at 4°C. The resin was then collected on a column and a gradient of 0-8 M guanidine hydrochloride passed through this column within a 4 hr period using constant levels of the buffered solution. The tubes were monitored by absorption at 300 mμ

(continued on next page)

eluted by 4 M guanidine hydrochloride conditions. The high absorption observed in Figure 19 under conditions of a 5-7 M
guanidine hydrochloride extract which yields only small amounts
of protein is due primarily to the elution of ribonucleoprotein
as well as small amounts (about 10% of the total) of DNA. No
DNA was extracted from this resin with up to 5 M guanidine hydrochloride treatment. Above this level, DNA began to dissociate from the resin, the amount of which was proportional to
the length of exposure. Subsequent experiments were performed
in which stepwise extraction of the chromatin-cellulose resin
was performed. After the extraction with increasing unit molarity levels of guanidine hydrochloride, samples of the resin
were removed, washed thoroughly, and subjected to the hormone
binding assay. As shown in Figure 20, the removal of all of
the histone and most of the acidic proteins from the chromatin
by 4 M guanidine hydrochloride failed to unmask significantly
the high affinity class of binding sites in the oviduct chromatin. These results support those shown previously in which
the removal of all of the histones did not unmask any more of
this class of binding sites. Further extraction of the chromatin-cellulose with 5 M and 6 M guanidine hydrochloride did unmask many of the high affinity binding sites. However, the 7 M
guanidine hydrochloride extraction greatly reduced the number
of high affinity binding sites remaining on the DNA.

Figure 21 shows the results of another experiment in which
the oviduct chromatin resin extracted with selected levels of
the guanidine hydrochloride was subjected to the binding assay
under increasing levels of the receptor. As can be seen in
this figure, the extraction of the chromatin with 5 or 6 M

Fig. 19 continued. (260-280 mμ gave too high an absorption,
consequently, 300 mμ had to be used in order to fully detect
each of the peaks). The fractions were also monitored for conductivity as well as refractive index, and the gradient level
of guanidine hydrochloride plotted. The fractions were collected according to their elution with each unit of concentration of guanidine hydrochloride (e.g., 1 M, 2 M, 3 M, etc).
Pooled samples were then dialyzed thoroughly versus water and
lyophilized. The lyophilized materials were resuspended in
a small volume of water, homogenized in a Teflon pestle glass
homogenizer and assayed for protein by Lowry. Total recoverable protein in each extract was estimated to be 80-90% of the
total protein on the column as chromatin-cellulose. The amount
of protein contained in each fraction is shown in the lower
graph.

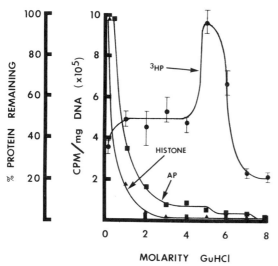

Fig. 20. The binding of ^{3}H-progesterone-receptor complex to hen oviduct chromatin preparations using the cellulose method under 0.15 M KCl conditions. In another experiment, aliquots of chromatin-cellulose were washed twice in 20 volumes of solution which contains 0.1 M HSETOH + 0.05 sodium bisulfate + 0.01 M Tris HCl (pH 8.5) and a specified molarity of guanidine hydrochloride. These resins were then washed in dilute Tris-EDTA buffer several times, frozen and lyophilized. Twenty to 50 mg amounts of the resin were then assayed for protein by resuspending the resin in the Tris-EDTA buffer, allowing it to hydrate several hours with gentle mixing, and assaying for histone using a 0.4 N H_2SO_4 extract with subsequent analysis of the acidic proteins (AP) with a 0.1 N NaOH extract (12,14). These guanidine-treated resins were tested for their capacity for binding the ^{3}H-progesterone-receptor complex using 20-25 μg of DNA (bound to the cellulose) together with 300 μl of the hormone-receptor complex solution. The binding assay was performed essentially as described in Methods, and the counts per mg DNA calculated. The ranges of 3 replicates of analysis for each assay of the hormone binding are shown.

Fig. 21. (Figure on next page) Binding of the ^{3}H-progesterone-receptor complex to hen oviduct chromatin fractions using the cellulose method under 0.15 M KCl conditions. The untreated chromatin-cellulose or the guanidine-treated chromatin-cellulose preparations were prepared essentially as described in the legend of Fig. 20. In each assay, 20-25 μg of DNA (as cellulose) were mixed together with increasing levels of the progesterone-receptor complex and incubated for 90 min at 4°C. This binding assay is essentially that described in Methods. The ranges of three replicates of analysis for each assay are shown.

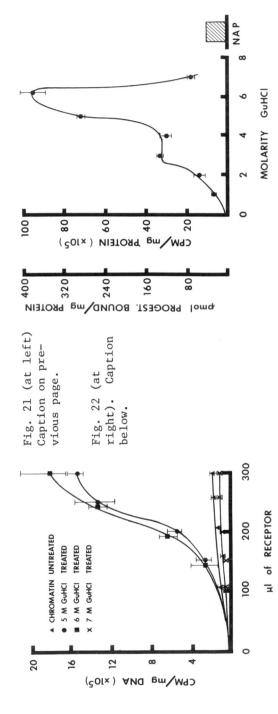

Fig. 21 (at left). Caption on previous page.

Fig. 22 (at right). Caption below.

Fig. 22 (at right). Binding of ^{3}H-progesterone-receptor complex to oviduct chromatin preparations using the cellulose method under 0.15 M KCl conditions. This binding assay is essentially that described in the legend of Fig. 21 and in Methods, with the exception that the radioactivity (CPM) was calculated per mg protein remaining bound to the DNA in place of the standard CPM/mg DNA. In addition, the pmol of the progesterone bound per mg protein was also calculated using the specific activity of the labeled hormone as described in Methods. The guanidine hydrochloride-treated chromatin-cellulose preparations were prepared essentially as described in the legends of Figures 19 and 20. The binding to the nucleoacidic protein (NAP) is shown for comparison. The ranges of three replicates of analysis for each assay are shown. The level of protein in each of the residual extracted chromatin-cellulose preparations was estimated from the data obtained in Fig. 20.

182

guanidine hydrochloride unmasked a large number (10-20 fold) of the high affinity binding sites. Further removal of the remaining small amounts of protein using 7 M guanidine hydrochloride extraction showed the complete abolishment of the binding capacity of the residual material. Little DNA was lost from these resins during these treatments. It appears that an unmasking of more high affinity binding sites occurs between 4 to 6 guanidine hydrochloride extraction and that these acceptors are then dissociated from the DNA by 7.0 M guanidine hydrochloride treatment.

Using this extraction procedure, the extent of purification of the acceptor belonging to the highest affinity class of binding sites was calculated using the amount of bound hormone per mass of protein. As shown in Figure 22, the treatment of chromatin with increasing levels of guanidine hydrochloride resulted in an immediate increase in the specific activity of the residual protein in binding the hormone-receptor complex up to 6 M. In this one procedure alone, almost a 100-fold purification was achieved. The extraction of the chromatin from 6 M to 7 M guanidine hydrochloride caused an immediate and marked decline in the binding capacity of the residual material on the resin. This fraction was found to contain protein and some RNA. The specific activity of the protein obtained with the nucleoacidic protein which was isolated using high salt-urea is also shown for comparison purposes. By this technique, the purification of the acceptor molecules belonging to the highest affinity class of binding site was 100-fold beginning with whole chromatin, several hundred-fold beginning with whole nuclei, and 98,000-fold beginning with whole tissue. The acceptor molecule belonging to the highest affinity binding sites in oviduct chromatin which bind the progesterone-receptor complex is not DNA, but probably a proteinaceous macromolecule belonging to the acidic (nonhistone) chromosomal protein.

DISCUSSION

The suggestion that the acceptor for the progesterone-receptor complex in the chick or hen oviduct is protein is supported by these studies. The evidence presented here supports earlier work in this laboratory implicating the acidic proteins as hormone "acceptor" (2,3,4,9). The acidic protein fraction designated AP_3 appears to contain the acceptors belonging to the highest affinity class of binding sites for the progesterone-receptor complex in the hen oviduct, as it did in the chick oviduct system described earlier (3,4). The subnuclear localization of this acceptor belonging to the highest affinity binding site is unknown at present. As depicted in Figure 1,

this acceptor protein may be associated with the DNA together
with the nuclear envelope or could be localized somewhere with
the DNA in the nucleoplasm. The evidence presented here for
the presence of multiple binding sites for the progesterone-
receptor complex in avian oviduct nuclei under <u>in vitro</u> binding
conditions is the first to be reported with sufficient data to
the knowledge of the authors. The higher affinity class of
binding sites appears to be tissue-specific, in that they are
not found in erythrocyte and spleen nuclei or chromatin. These
higher affinity binding sites appear to be associated with the
acidic (nonhistone) proteins of chromatin.

The majority, but not all, of the high affinity binding
sites are masked in the target cell chromatin. Practically all
of these binding sites are masked in the chromatin of the nontarget cell. Histones, the impressive maskers of genes with
respect to their transcription, do not appear to be involved in
the masking of these high affinity binding sites for the progesterone-receptor complex. It appears that other fractions
of acidic chromatin protein are involved in the masking of these
acceptor molecules. The possibility that some fractions of
acidic proteins actually mask and cover over other fractions
of acidic proteins lends support to the intricacies of the
ultrastructure and arrangement of components in chromatin. It
is proposed that only the highest affinity class of binding
sites identified in these experiments represent the real nuclear
binding site for progesterone in the hen oviduct nucleus. The
reasons for this are three-fold. Firstly, it is only logical
to assume that the true binding site for steroid-receptor complexes in the nucleus would be one of very high affinity, certainly much higher than nonspecific binding sites. Secondly,
under the ionic conditions of a cell and even higher ionic
conditions proposed to be the environment in the cell nucleus,
only the highest affinity class of binding sites survive, as
shown in the results of this paper. Lastly, these highest
affinity binding sites on chromatin were found to be tissue-
specific, found only in the target cells for progesterone and
not in nontarget cells. It should be pointed out that the lower
affinity classes of binding sites observed in the <u>in vitro</u>
binding assays in the hen oviduct nucleus with the progesterone-
receptor complex may be biologically important if they are
bound by the hormone-receptor complex under situations where
there may be 20,000 or more hormone-receptor molecules, all of
which can enter into the nucleus. Alternatively, these lower
affinity sites may be important under conditions of pharmaco-
logical doses of steroid hormones under which genes which code
for production of drug metabolizing enzymes (mixed function
oxidases) are activated to reduce this poisonous level of hor-

mone in the cell. The result of the interaction of steroid-receptor complex with the genetic material is extremely important, in that the result of this interaction is an alteration of RNA polymerase activity and of RNA synthesis in general. Studies are underway to demonstrate that the in vivo bound progesterone will serve to compete with the subsequently in vitro bound labeled progesterone for the acceptor. Similar investigations with the estrogen acceptor are planned.

Acknowledgment

This work was supported by grants HD 08441 and CA 14920 and the Mayo Foundation.

References

1. H. Busch. "Histones and Other Nuclear Proteins", Academic Press, New York (1965), p. 227.

2. T. C. Spelsberg, A. W. Steggles and B. W. O'Malley. J. Biol. Chem. 246:4188 (1971).

3. T. C. Spelsberg, A. W. Steggles, F. Chytil and B. W. O'Malley. J. Biol. Chem. 247:1368 (1972).

4. T. C. Spelsberg. In: "Acidic Proteins of the Nucleus". I. L. Cameron and J. R. Jeter, Jr., eds. Academic Press, New York (1974), p. 247.

5. R. J. B. King, J. Gordon and A. W. Steggles. Biochem. J. 114:649 (1969).

6. G. E. Swaneck, L. L. H. Chu and I. S. Edelman. J. Biol. Chem. 245:5382 (1970).

7. E. V. Jensen and E. R. DeSombre. Ann. Rev. Biochem. 41: 203 (1972).

8. B. W. O'Malley, D. O. Toft and M. R. Sherman. J. Biol. Chem. 246:117 (1971).

9. B. W. O'Malley, T. C. Spelsberg, W. T. Schrader, F. Chytil and A. W. Steggles. Nature 235:141 (1972).

10. T. C. Spelsberg, L. S. Hnilica and A. T. Ansevin. Biochim. Biophys. Acta 228:550 (1971).

11. J. M. Neelin. Can. J. Biochem. 46:241 (1968).

12. O. H. Lowry, N. J. Rosebrough, A. L. Farr and R. J. Randall. J. Biol. Chem. 193:265 (1951).

13. W. T. Schrader and B. W. O'Malley. J. Biol. Chem. 247:51 (1972).

14. K. Burton. Biochem. J. 62:315 (1956).

15. T. C. Spelsberg. In preparation.

16. R. M. Litman. J. Biol. Chem. 243:6222 (1968).

17. T. C. Spelsberg and E. Stake. In preparation.

18. R. E. Buller, D. O. Toft, W. T. Schrader and B. W. O'Malley. J. Biol. Chem. 250:801 (1975).

19. R. E. Buller, W. T. Schrader and B. W. O'Malley. J. Biol. Chem. 250:809 (1975).

20. T. C. Spelsberg. Unpublished.

21. J. D. Duerkson and B. J. McCarthy. Biochemistry 10:1471 (1971).

22. E. C. Murphy, S. H. Hall, J. Shepherd and R. S. Webster. Biochemistry 12:3843 (1973).

23. S. Panyim and R. Chalkley. Arch. Biochem. Biophys. 130:337 (1969).

24. E. Wilson and T. C. Spelsberg. Biochim. Biophys. Acta 322:145 (1973).

CHROMATIN PROTEINS: ELECTROPHORETIC, IMMUNOLOGICAL AND METABOLIC CHARACTERISTICS

Harris Busch, Raghuveera Ballal, Rose K. Busch,
Edward Ezrailson, Ira L. Goldknopf, Mark O. J. Olson,
Archie W. Prestayko, Charles W. Taylor and Lynn C. Yeoman
Nuclear Protein and Tumor By-Products Laboratories
Department of Pharmacology
Baylor College of Medicine
Houston, Texas 77025

Abstract

Two-dimensional polyacrylamide gel electrophoresis shows that in nuclei of Novikoff hepatoma ascites cells there are approximately 75 proteins in the chromatin fraction soluble in 3 M NaCl-7 M urea. Dialysis of this fraction to an ionic strength of 0.14 produces a soluble fraction and a precipitate. The proteins in the soluble fraction have been reported to be active in gene control. Antibodies to the soluble fraction distributed diffusely throughout the nucleus and antibodies to the precipitate localized primarily in the nucleolus and the nuclear ribonucleoprotein network. Autoradiography of dried two-dimensional gels after labeling of the nuclear proteins in vivo with ^{3}H-L-leucine demonstrated that of the major proteins in the soluble fraction, proteins C6, C14 and C13 were most highly labeled. Some of these proteins have now been fractionated by chromatography on DNA-polyacrylamide columns. The nucleolar proteins differ from the extranucleolar in antigenicity and labeling patterns. Specific localization to the nucleolar chromatin fraction has now been demonstrated for at least one phosphoprotein, C18.

The evidence for an important role of NHP (nonhistone nuclear proteins) in gene control, the subject of recent reviews (1-3), includes findings of the multiplicity of species of these molecules (2,4,5) as well as alterations of gene readouts in their presence (6-10). It was also shown (1-3,10) that addition of protein fractions from specific tissues to chromatin fractions of other tissues produces alterations of gene readouts. Recently, this method has demonstrated production of mRNA$_{glob}$ in tissues that are not erythropoietic (10-12). This work supplements hybridization-competition studies that show

187

products of a given tissue can be produced in chromatin of
other tissues by addition of the appropriate NHP (6-12).

The evidence that there is an increased readout of $mRNA_{oval}$
as a result of interactions between specific cytoplasmic proteins or specific "receptor proteins" (10) and chromatin (or
DNA) has resulted in a search for the precise mechanisms involved in this and other gene activation reactions. Unfortunately, the purity of the products involved has not yet been
established. Despite the suggestive evidence that definitive
interactions between some protein fractions and chromatin cause
the production of specific mRNA species, no specific example
of interaction has yet been demonstrated between a highly purified "gene activator" protein and chromatin that produces
a single species of mRNA.

On the other hand, the data that support the concept of a
special role for NHP in gene activation have been subject to a
great deal of criticism. This results from the conditions under which many of the studies were carried out and the claim
there is production of specific mRNA. For example, it has been
pointed out that the RNA polymerases generally used could not
have had high fidelity in terms of reading the correct DNA
strands (13) and further that the RNA produced was only of low
Cot (highly reiterated) rather than "unique sequence" or specific mRNA.

Moreover, the large number of nuclear proteins found in
studies in this and other laboratories (1,2) have not yet been
categorized with respect to function but it is certain that
many will be shown to have enzymatic and structural roles.
Accordingly, there is a need for further understanding of the
molecular species of chromatin NHP and their possible roles in
gene control.

Evidence has been presented that gene control activities
are not generally distributed throughout the group of nuclear
proteins but are extracted with salt solutions of high ionic
strength (6-9,14-18). However, Holoubek and Fujitani (19,20)
and Kostraba et al (21) have indicated that the proteins extracted from chromatin with 0.35 M NaCl and 0.60 M NaCl are
similar to those extracted with the higher salt concentrations
as shown by one-dimensional polyacrylamide gel electrophoretic
analysis. In preliminary studies of Kostraba et al (21) it was
also found that this fraction exhibited a significant activity
in stimulation of gene readouts. The present studies were
designed to analyze the selectivity of the extraction with high
salt by two-dimensional electrophoresis of the proteins and also to provide immunological and metabolic comparisons of these

proteins with those of other fractions (22,23).

MATERIALS AND METHODS

Isolation of Nuclei and Preparation of Chromatin. Nuclei
were isolated by the citric acid method (22,23) from NKM (0.13
M NaCl, 0.005 M KCl, 0.008 M $MgCl_2$) washed Novikoff hepatoma
ascites cells transplanted in normal male albino rats (Holtzman
Co., Madison, Wis.). The nuclei were purified in sucrose solutions containing 0.1 mM PMSF (phenylmethyl sulfonyl fluoride).
Similar preparations of liver nuclei were obtained (22). Chromatin was prepared from these nuclei by the method of Marushige
and Bonner (24) and the extractions were made with 0.35 M NaCl,
0.60 M NaCl or 3 M NaCl-7 M urea containing 0.01 M Tris, pH 8,
and 0.1 mM PMSF, either directly or successively.

Fractions Obtained from the Chromatin. The fractions obtained from these extractions were the soluble proteins and
an insoluble residue. Some of the fractions, such as those
extracted with either 2 M NaCl or 3 M NaCl-7 M urea were dialyzed against sufficient 0.01 M Tris buffer to reduce the ionic
strength to 0.14 to precipitate the DNA-histone and associated
proteins.

Preparation of 0.35 M NaCl Soluble Chromatin Proteins.
Chromatin from rat liver citric acid nuclei was extracted three
times with buffer containing 0.01 M Tris-HCl, pH 8.0, 0.35 M
NaCl and 0.1 mM PMSF at a ratio of 10 ml of extraction solution
per gram of nuclei (25). The pooled extract was centrifuged
at 142,000 x g for 24 hours.

The soluble proteins were concentrated by an Amicon apparatus fitted with a UM-10 filter or by precipitation with ethanol (3:1, v:v), taken up in a solution containing 0.9 M acetic
cid, 10 M urea and 1% β-mercaptoethanol and then dialyzed
against two changes of the same buffer for 16 hours or against
0.05 M Tris-HCl, pH 7.8, for chromatography on a DNA-polyacrylamide column. Protein was determined by the Lowry method (26)
with bovine serum albumin as a standard (Miles Lab., Kankakee,
Ill).

Two-Dimensional Gel Electrophoresis of the Proteins. The
samples were analyzed by the two-dimensional electrophoretic
procedure described previously (27,28) with the modification
that the gel used for the first dimension contained 6% polyacrylamide and the second usually contained 8% polyacrylamide
(29).

Preparation of Antibodies. Protein samples (10-50 mg) were injected intradermally and intramuscularly in 3 weekly doses into New Zealand white rabbits in 1 ml of Freund's adjuvant diluted 1:2 with 0.15 M NaCl. The blood was collected from the ear veins 7-10 days after the third inoculation (30). The procedure used for detecting immunofluorescence was a modification of that of Hilgers et al (31).

Labeling of the Proteins. Three hours after each tumor-bearing rat was injected intraperitoneally with 5 mc of ^{3}H-L-leucine, the rats were decapitated and the cells were collected from the ascites fluid. The nuclei, chromatin and other products were isolated as indicated above. For autoradiography the gels were impregnated with PPO (32), dried and placed on Kodak RP-Royal X-omat film at -70° for 12-24 hours.

DNA Polyacrylamide Chromatography. DNA polyacrylamide columns were prepared according to Cavalieri and Carroll (33). Rat liver DNA was prepared by a modified Marmur (34) procedure according to Sitz et al (35). DNA polyacrylamide was prepared with a composition of 5% polyacrylamide, 0.5% agarose and 0.01% native rat liver DNA. The DNA polyacrylamide was pressed through a stainless steel wire mesh (#80), washed six times with 5 volumes of 0.05 M Tris-HCl, pH 7.8, buffer and packed in a 1 x 30 cm column. A protein sample of 10 to 15 mg was applied to the column and the column was washed with 200 ml buffer. The DNA-bound protein was eluted with 2 M KCl. Proteins from the column were analyzed by two-dimensional polyacrylamide gel electrophoresis.

RESULTS

Comparison of Nuclear Fractions Obtained with Salt Extractions. Many procedures are currently in use to extract various types of soluble and difficultly soluble proteins from the nucleus and nucleolus. Of the extraction solutions used to solubilize chromatin proteins and DNA, 3 M NaCl-7 M urea is probably more efficient than the other saline solutions used. This extraction is usually preceded by initial nuclear extractions with 0.075 M NaCl-0.025 M Na EDTA, pH 7.4-8.0, and intermediate extraction with 0.01 M Tris, pH 7.4-8.0. Addition of the 0.1 mM PMSF effectively prevents proteolysis; in these experiments NaHSO$_3$ was shown to be ineffective (7,8,25).

Figure 1 shows that the proteins soluble in 3 M NaCl-7 M urea include histones (A17-A19), the histone-like protein A24 and many NHP in the B and C region. An alternative procedure for obtaining NHP from "dehistonized" chromatin is the Wilson-

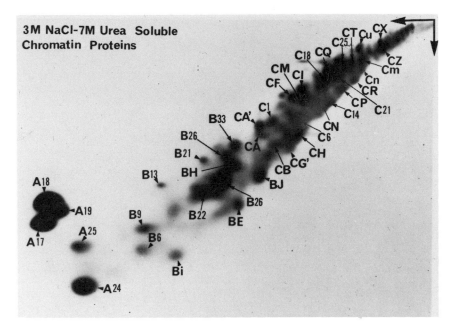

**3M NaCl-7M Urea Soluble
Chromatin Proteins**

Fig. 1.   Two-dimensional gel electrophoretic patterns of proteins extracted from Novikoff hepatoma chromatin with 3 M NaCl-7 M urea.   Histones are present in the extract but most of the other proteins are identical to those obtained with older procedures (28).   The samples were prepared as described by Orrick et al (27) and Yeoman et al (28) and the "6-8" gels were run as described by Busch et al (29).

Spelsberg procedure (36).   In this procedure, chromatin extracted first with EDTA and 0.01 M Tris (pH 8.0) is then extracted with 0.4 N H_2SO_4 to remove histones and small amounts of the nonhistone acid soluble proteins.   The remainder of the proteins in the chromatin are extracted by treatment with DNase I and subsequent precipitation with 0.4 N $HClO_4$.   The corresponding profile of these proteins on two-dimensional gels is shown in Figure 2 which indicates that most of the proteins of the 3 M NaCl-7 M urea extract are present in these samples.   For example, proteins B22, 24, 26, H and 33 are readily seen in both patterns.   Proteins CA, CA', C1 and CC are also easily visualized.   In addition, the proteins C14, CP, C18, CQ and C21 are also well separated as are proteins CR, CT, CU and Cm.

Proteins in 0.35 M NaCl Extractions of Chromatin.   One of the important problems in isolation of nuclear proteins has been the high concentration of salt and urea in the various ex-

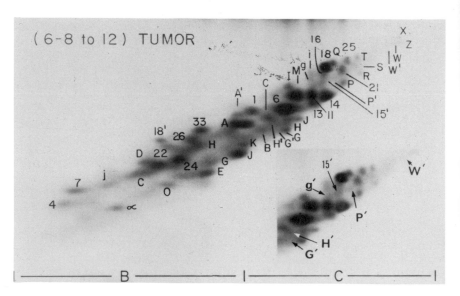

Fig. 2. Legend on next page.

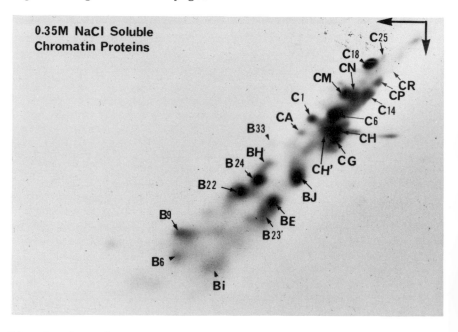

Fig. 3. Legend on next page.

traction solutions employed. These solutes, which are denaturing, have complicated the task of separation and isolation of the various proteins. Recently, Johns and Forrester (37), Fujitani and Holoubek (20) and Kostraba et al (21) have reported that the fraction extracted with 0.35 M NaCl has similar proteins to those present in extracts with more concentrated salt solutions. To determine whether the proteins extracted from chromatin were similar to those extracted with higher salt concentrations (19-21,37), a two-dimensional gel electrophoretic analysis was carried out (Fig. 3). Most of the proteins in this extract are present in the extract obtained with 3 M NaCl-7 M urea (Fig. 1). Such extracts contain large amounts of proteins B22, 24 and H as well as proteins BE, BJ, B9, B6, Bi and B23'. In the C region C1, CG, CH, C6, CM, CN, C14, C18 and CP are present in the 0.35 M NaCl extract (Fig. 3).

These results support the concept that there is compartmentation of the nuclear proteins and that a pool of these proteins is not tightly associated with the DNA but may equilibrate with it. For analyses of a large number of these proteins, extraction with 0.35 M NaCl is a good operational simplification over extraction with 3 M NaCl-7 M urea.

It is not surprising that many of these same proteins are present in the fraction referred to by Kostraba and Wang (17, 18) as the "soluble" chromatin fraction (Fig. 4). In our laboratory, this fraction is obtained by initial extracts with 3 M NaCl-7 M urea followed by dialysis of the supernatant against sufficient water to reduce the ionic strength of the salt solution to 0.14.

Which, if any, of these proteins are "gene control" proteins? It has been suggested that proteins present in large quantities either serve roles as specific enzymes or as "carrier" proteins. For example, B24 has been shown in our laboratory (Prestayko et al, in preparation) to be the "Samarina"

Fig. 2. Two-dimensional gel electrophoretic patterns of proteins extracted by treatment of "dehistonized" Novikoff hepatoma chromatin with DNase I followed by precipitation of the proteins with 0.4 N HClO$_4$. The gels were run as described for Figure 1.

Fig. 3. Two-dimensional electrophoretic gel pattern of proteins extracted from Novikoff hepatoma chromatin with 0.35 M NaCl. The remarkable similarity of these proteins to those observed in Figures 1 and 2 suggests that such proteins are in equilibration with those more firmly bound to the chromatin.

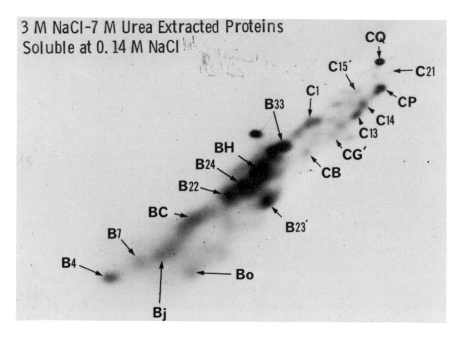

Fig. 4. Two-dimensional gel electrophoretic pattern of proteins extracted from chromatin with 3 M NaCl-7 M urea and remaining soluble after dialysis of the extract to an ionic strength of 0.14 M. This protein fraction has been suggested to contain "gene control proteins" (18). In agreement with earlier suggestions of Holoubek and Fujitani (19) and Kostraba et al (21), these proteins exhibit migration characteristics that are extremely similar to those shown in Figure 1 for extraction of chromatin made with more concentrated salt solutions.

protein or "informatin" (38). Whether B22, BH and B33 may serve similar roles is not clear at the moment. Proteins Cl8, Cg' and C21 have been shown to be predominantly in the nucleolus. It is not clear whether these proteins serve roles as transport or carrier elements for RNP particles or other nucleolar functions.

Antibodies to Chromatin and Nucleolar Proteins. To ascertain whether there were immunological differences in the chromatin and nucleolar proteins, antibodies were prepared in rab-

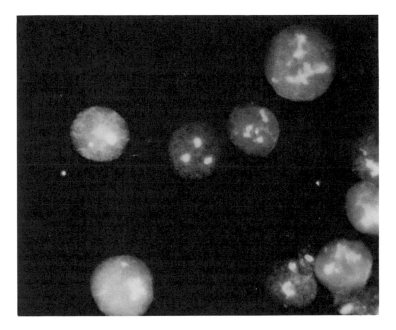

Fig. 5A (above) & 5B (below). Legends on next page.

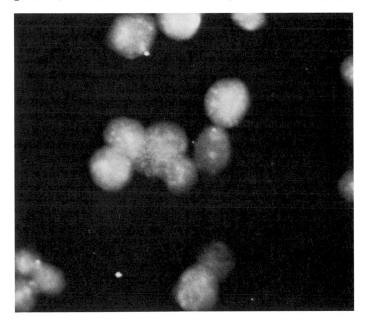

bits and their interactions with nuclei were studied by immunofluorescence (Fig. 5,6). Antinucleolar antibodies (Fig. 5A)
exhibited a high order of specificity in localization to nucleoli (30), and antibodies to the Kostraba-Wang soluble fraction
distributed generally throughout the nucleus (Fig. 5B) as a
whole without preferential staining of nucleoli. The multiple

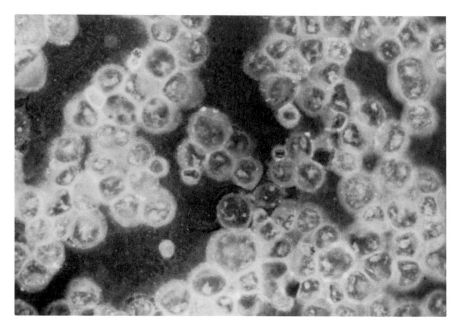

Fig. 6A. Immunofluorescence produced in cells after exposure to antinucleolar antibodies.

small regions of local density may reflect binding to some
chromocenters. Antibodies to the Kostraba-Wang (17,18) fraction insoluble at an ionic strength of 0.14 were localized primarily to the nucleolus and in the fibers of the nuclear ribonucleoprotein network. The localization of the antibodies
strongly resembles stained pictures of the network fibers that
were described in earlier reports (22,30).

Fig. 5 (A) Immunofluorescence produced in nuclei isolated
by the citric acid method on exposure to antinucleolar antibodies (30). (B) Immunofluorescence produced in nuclei isolated by the citric acid method with antibodies to the chromatin fraction soluble in 3 M NaCl-7 M urea after reduction
of the ionic strength to 0.15. Antibody localization was
detected by indirect immunofluorescence, i.e., by staining with
fluorescein labeled goat antirabbit antibodies.

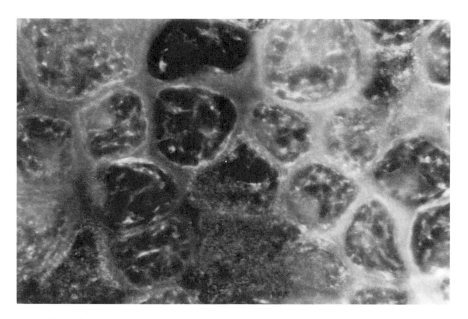

Fig. 6B. Immunofluorescence produced in cells after exposure to antichromatin antibodies.

Turnover of Various Chromatin Proteins. Earlier studies
have indicated that there are differences in turnover rates of
histone and nonhistone nuclear proteins. However, detailed
studies have not yet been made on turnover rates of any of the
individual NHP. Accordingly, studies were initiated on labeling of various proteins with either ^{32}P or ^{3}H-leucine, the former to label phosphoserine and phosphothreonine residues and
the latter to label leucines in the amino acid chain. For
studies with ^{32}P, the radioactive phosphate was injected into
the ascites fluid of tumor bearing rats or was incubated with
ascites cells under tissue culture conditions. Many proteins
were found to be phosphorylated, although relatively few contained significant amounts of ^{32}P on a molar basis. Interestingly, the labeling patterns for nucleolar and extranucleolar
proteins differed significantly (Fig. 7-9).

The most remarkable finding obtained thus far is that protein C18 was highly labeled in the nucleolus and contained
approximately 40-50% of the total ^{32}P of the nucleolar proteins.
On the other hand, the extranucleolar chromatin fraction contained 6-8 phosphoproteins which appeared as dense spots on
the autoradiograph. These proteins contain C18 as well, but

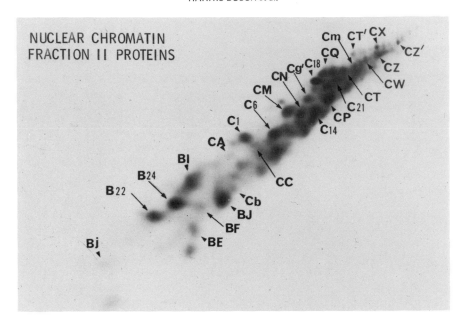

Fig. 7 A (above). Stained gel of nuclear chromatin II proteins separated by two-dimensional electrophoresis as noted in Figure 1. B (below). Autoradiograph of phosphoproteins in the gel shown in Figure 7A.

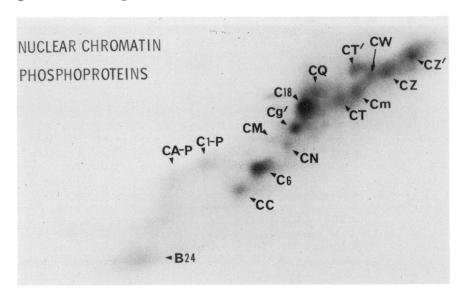

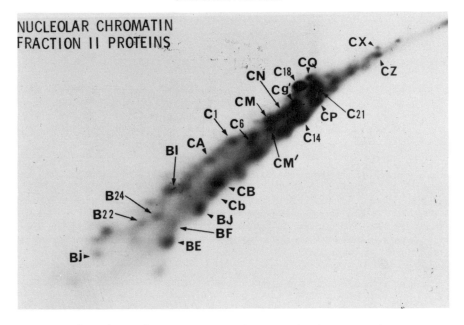

Fig. 8 A (above). Stained gel of nucleolar proteins separated by two-dimensional electrophoresis as noted in Figure 1. B (below). Autoradiograph of nucleolar phosphoproteins in the gel shown in Figure 8A.

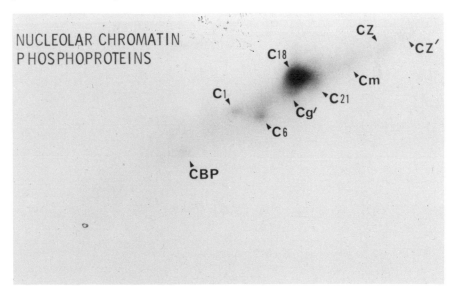

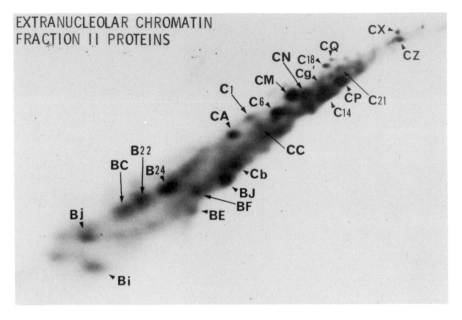

Fig. 9 A (above). Stained gel of extranucleolar proteins separated by two-dimensional electrophoresis as noted in Figure 1. B (below). Autoradiograph of extranucleolar phosphoproteins in the gel shown in Figure 9A.

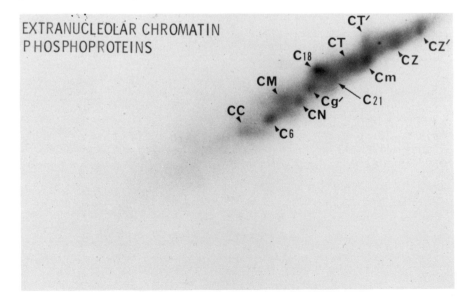

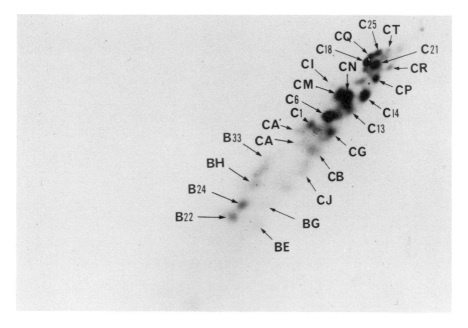

Fig. 10A. Comparative staining and autoradiographic patterns of a two-dimensional gel electrophoretic pattern of chromatin II proteins of Novikoff hepatoma ascites cells. The
stained pattern is comparable to that of Figure 1B.

the concentration as determined for both the staining and ^{32}P
content was significantly lower than that of the nucleolus.

Labeling with ^{3}H-Leucine. Inasmuch as earlier studies from
this laboratory established that the bulk of the ^{32}P incorporated into nuclear proteins was present in ester linkage in
threonine or serine, it was clear that this method could not be
used for estimation of turnover or synthesis of amino acid
chain of the proteins. Accordingly, high specific activity ^{3}H-
L-leucine was either injected in tumor bearing rats or incubated in tissue culture systems to label the proteins. Figure
10 shows autoradiographs and stained patterns for 2D gels obtained for the "Chromatin Fraction II" (28,29) that contains
most of the nonhistone chromatin proteins. It is of special
interest that the densely stained proteins C18, CQ, C25, CM and
CN contain very little isotope by comparison with the densely
stained proteins C6 and C14. C13, which is a less densely
stained protein, contains significant amounts of tritium, as
does CG.

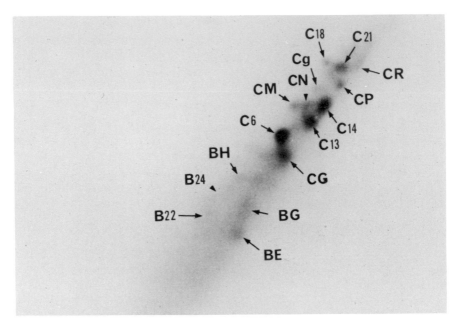

Fig. 10B. The autoradiogram indicates marked differences in the in vivo labeling (3 hours, 5 mC, ^{3}H-L-leucine) of various proteins. The autoradiographic spots were more dense for proteins C6 and C14 and much less dense for proteins of high density by staining.

In the B region, spots B22, B24 and BH, which are relatively densely stained in most chromatin protein patterns, contain relatively little isotope. It is of interest that some of the nuclear proteins, particularly C6, C13 and C14, have high rates of labeling with tritiated leucine while others have much lower rates of labeling. The functions of these proteins are not understood at the present time, but it is clear that there are remarkable differences in their turnover. It seems possible, as was suggested by Yu and Feigelson (39) that some nuclear proteins have very short half-lives. Some proteins with short half-lives may be involved in gene control or nucleocytoplasmic interactions. In the latter case, nuclear turnover would be a reflection of overall cytoplasmic synthetic reactions and rates of transport and binding in the nucleus.

Protein B24, which has been found in other studies to contain the "informatin" of Samarina et al (38) was labeled with

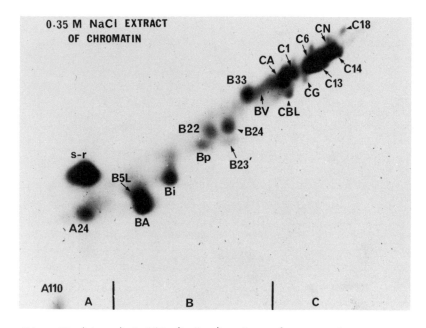

Fig. 11 (above) & 12A (below). Legends on next page.

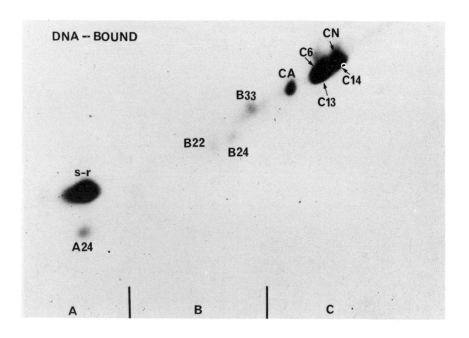

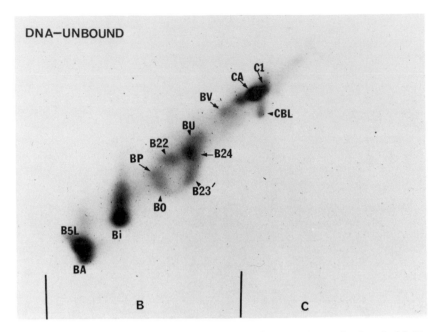

Fig. 12B. 2-D electrophoresis of proteins of the 0.35 M extract of chromatin that did not bind to the DNA-polyacrylamide column and were eluted in the void fraction. The gel running conditions are the same as in Figure 11.

Fig. 11. 2-D electrophoresis of proteins extracted with 0.35 M NaCl from normal rat liver chromatin prepared from citric acid nuclei. Proteins were separated on a 6% polyacrylamide gel in the first dimension and on an 8% gel in the second dimension.

Fig. 12A. 2-D electrophoresis of proteins of the 0.35 M NaCl extract of chromatin eluted from the DNA-polyacrylamide column with 0.05 M Tris-HCl buffer, pH 7.8, containing 2 M KCl. The gel running conditions are the same as in Figure 11.

[3]H-leucine only to a very small extent; such carrier or "shut-
tling" proteins for RNA may, in fact, turn over at very slow
rates in these tumors. Interestingly, it was highly labeled
with [32]P. A great deal needs to be learned about specific activities of these proteins and their individual turnover. For
such purposes, kinetic studies on specific activity are required. At present, satisfactory methods for determining specific activities of proteins in gels are lacking. Extraction
of the proteins from the gels was effected earlier by electrophoresis. Periodate oxidation of the gels formed in the presence of diallyltartaramide (a cross-linking agent used in place
of bisacrylamide) has the advantage that this cross-linking
agent contains a cis-diol that is readily cleaved by periodate
oxidation. Accordingly, it is possible to extract the proteins
from the long polyacrylamide fibers without solubilizing the
bulk of the gel.

DNA-Binding Proteins. To determine which of the nuclear
proteins exhibits DNA binding activity, studies are currently
being made on DNA polyacrylamide columns prepared by the method
of Cavalieri and Carroll (33). In the present studies, liver
chromatin proteins extractable with 0.35 M NaCl were used. Of
the proteins applied to the column, 10-13% were bound to DNA
and eluted with 2 M KCl. Columns prepared without DNA did not
retain any protein. A two-dimensional gel electrophoretic
pattern of the total protein in the 0.35 M NaCl extract of
chromatin is shown in Figure 11. Of the 20 spots, 12-14 are
major protein spots. The smaller proteins in this fraction
are A24 and BA with molecular weights of about 20,000-25,000
daltons. The largest proteins are of about 70,000-80,000
daltons.

When the proteins were passed through the DNA polyacrylamide column, 5 proteins (AS, Ar, C6, CN, C14) were quantitatively bound to the column while spots A24, B24, B33 and CA
were bound to lesser extents (Fig. 12A). The gel electrophoretic pattern of the proteins which pass through the column in
the void fraction is shown in Figure 12B. Spots BA and Bi
are quantitatively recovered in this fraction. Some proteins
with very low intensities in Figure 11 were found to be present
in much larger amounts in this fraction. Thus, relatively
few proteins of both high and low molecular weight bind stongly
to DNA. The possibility exists that these proteins of the
chromatin fraction are involved in maintenance of chromatin
structure or serve specific functional roles. Studies on possible functions of these DNA-binding proteins will be carried
out in reconstitution systems.

DISCUSSION

The initial goal of these studies was to determine whether or not similar proteins were present in the Chromatin Fraction II (28) and the fraction extracted from chromatin with 3 M NaCl-7 M urea. The two-dimensional gels indicate the similarity of these fractions. They also show similar proteins are extracted with 0.35 M NaCl (a more optimal extractant for such studies) and the proteins soluble at an ionic strength of 0.14 after initial dissociation and extraction with 3 M NaCl-7 M urea.

These results support prior studies which have indicated that chromatin proteins extracted at both low and high ionic strength contain many similar proteins (19-21,37). They also support the suggestion that many products in soluble chromatin fractions are in a dynamic equilibrium with similar proteins that are more tightly bound to chromatin. The exact dynamics of this equilibrium and its role in nuclear protein function must be assessed in future studies.

Although these studies indicate that specific proteins are present in substantial amounts in the chromatin fraction, they do not provide data on the possible functions of these proteins. These proteins may maintain structural states of the chromatin (including eu- and heterochromatin), mRNP (40, 41) or other RNP fractions or may have gene regulatory roles. For the latter, it seems quite likely that the protein concentrations required are substantially lower than those in which the spots are visible in the present studies.

Along with studies on labeling, the immunological studies have definitely indicated that different antigenic proteins are present in the nucleolar and chromatin fractions. The nucleolar antigens seem to be localized with a high degree of specificity. Their presence in the nucleoplasm in defined fibrous elements is in agreement with the prior suggestions that the nuclear ribonucleoprotein network contains nucleolar products in transit to the inner layer of the nuclear envelope. It remains to be seen which specific products are indeed responsible for the antigenicity. Such studies will require isolation of individual proteins as well as the possible synthesis of such proteins in in vitro systems.

As an approach to searching for proteins present in low concentration and in dynamic metabolic states as contrasted to the histones, autoradiographic studies have been initiated on proteins labeled with tritiated amino acids and ^{32}Pi. Fortunately, the level of labeling in vivo was sufficiently high to provide autoradiographic visualization of proteins in the C region such as C6, C14, C13, C21 as well as a number of other proteins. Interestingly, some proteins that are present in substantial amounts as indicated on the stained gels apparently have relatively low turnover rates. Presumably, such proteins may serve structural roles or may function as tightly bound repressors. The reasons why some chromatin proteins turn over rapidly can only be speculated on at the present time. Such proteins may be important in gene control, but they also may form complexes with messenger RNA (38,41,42) or other nuclear RNAs.

The methods employed and the findings of the present study offer interesting opportunities for more intensive investigations on nuclear and nucleolar proteins. Comparisons of proteins that are rapidly phosphorylated in the chromatin and the nucleolus like those rapidly labeled with the tritiated amino acids have shown that some proteins such as C6 have a rapid turnover of both the phosphate and leucine moieties. On the other hand, protein C18, which is highly phosphorylated in the nucleolar fraction, exhibits an apparently higher degree of phosphorylation than turnover of the amino acid chain. Currently, attempts are underway to isolate a variety of these proteins under non-denaturing conditions and to determine their functions in reconstituted systems. At present, an optimal reconstituted system appears to be the isolated nucleolus which has recently been shown for the first time to synthesize high molecular weight RNA under appropriately supplemented conditions.

Acknowledgment

The authors wish to express their appreciation to Michael R. Bannon, Alan R. Peaceman and Vincent Tam for their technical assistance in portions of this work.

TABLE I

EXTRACTANTS USED TO PREPARE
CHROMATIN AND NHP

I. Dilute Salt Solutions
    a) 0.075 M NaCl - 0.025 M Na EDTA
    b) 0.15 M NaCl - 0.015 M Na Citrate (pH 7-8)
    c) 0.01 M Tris (pH 7-8)

II. Concentrated Salt Solutions
    a) 0.35 M NaCl
    b) 0.6 M NaCl
    c) 2 M NaCl

III. Salt Urea Solutions
    a) 2 M NaCl - 5 M urea
    b) 3 M NaCl - 7 M urea

IV. Acids
    a) 0.25 M N HCl
    b) 0.4 N H_2SO_4
    c) 0.1 N $HClO_4$

TABLE II

TWO-DIMENSIONAL GEL OF NOVIKOFF
HEPATOMA CHROMATIN PROTEINS

	cpm
C14	2180
C13	1800
CP	240
CN-CM	2365
CG	1240
C6	1810
BJ	1060
BH	600
B24	730
B23	600

Table II. ^{3}H–Leucine in various spots of the two–dimensional Electrophoretogram of Chromatin Fraction II. The spots were excised from the dried gel, treated with 30% H_2O_2 for one hour at 25° and placed in liquid Aquasol^R and analyzed for isotope content in a liquid scintillation counter.

References

1. M. O. J. Olson, W. C. Starbuck and H. Busch. In: "The Molecular Biology of Cancer", H. Busch, ed. Academic Press, New York, p. 309 (1974).

2. M. O. J. Olson and H. Busch. In: "The Cell Nucleus, H. Busch, ed. Academic Press, New York, p. 212 (1974).

3. G. S. Stein, T. C. Spelsberg and L. J. Kleinsmith. Science 183:817 (1974).

4. W. J. Steele and H. Busch. Cancer Res. 23:1153 (1963).

5. H. Busch. "Histones and Other Nuclear Proteins", Academic Press, New York (1965).

6. J. Paul and R. S. Gilmour. Nature 210:992 (1966a); J. Mol. Biol. 16:242 (1966b).

7. T. C. Spelsberg, A. W. Steggles and B. W. O'Malley. Biochim. Biophys. Acta 254:129 (1971a); J. Biol. Chem. 246: 4188 (1971b).

8. T. C. Spelsberg and L. S. Hnilica. Biochim. Biophys. Acta 195:63 (1969).

9. N. C. Kostraba and T. Y. Wang. Biochim. Biophys. Acta 262:169 (1972).

10. B. W. O'Malley and A. R. Means. In: "The Cell Nucleus", H. Busch, ed. Academic Press, New York, p. 379 (1974).

11. J. Paul, R. S. Gilmour, N. Affara, K. Windass and P. Harrison. 9th FEBS Meeting, abstract #130 (1974).

12. R. Axel, H. Cedar and G. Felsenfeld. Proc. Nat. Acad. Sci. USA 70:2029 (1973).

13. T. Honjo and R. H. Reeder. Biochemistry 13:1896 (1974).

14. J. Paul and R. S. Gilmour. J. Mol. Biol. 34:305 (1968).

15. T. C. Spelsberg, L. S. Hnilica and A. T. Ansevin. Biochim. Biophys. Acta 228:550 (1971).

16. T. C. Spelsberg and L. S. Hnilica. Biochem. J. 120:435 (1970); Biochim. Biophys. Acta 228:202 (1971); Biochim. Biophys. Acta 228:212 (1971).

17. N. C. Kostraba and T. Y. Wang.   Cancer Res. 32:2348 (1972).

18. N. C. Kostraba and T. Y. Wang.   Exp. Cell Res. 80:291 (1973)

19. V. Holoubek and H. Fujitani.   Biochem. Biophys. Res. Commun. 54:1300 (1973).

20. H. Fujitani and V. Holoubek.   Experientia 30:474 (1974).

21. N. C. Kostraba, R. A. Montagna and T. Y. Wang.   In press.

22. H. Busch and K. Smetana.   "The Nucleolus", Academic Press, New York (1970).

23. C. W. Taylor, L. C. Yeoman, I. Daskal and H. Busch.   Exp. Cell Res. 82:215 (1973).

24. K. Marushige and J. Bonner.   J. Mol. Biol. 50:641 (1970).

25. N. R. Ballal, D. A. Goldberg and H. Busch.   Biochem. Biophys. Res. Commun., in press.

26. O. H. Lowry, N. J. Rosebrough, A. J. Farr and R. J. Randall. J. Biol. Chem. 193:265 (1951).

27. L. R. Orrick, M. O. J. Olson and H. Busch.   Proc. Nat. Acad. Sci. USA 70:1316 (1973).

28. L. C. Yeoman, C. W. Taylor, J. J. Jordan and H. Busch. Biochem. Biophys. Res. Commun. 53:1067 (1973).

29. G. I. Busch, L. C. Yeoman, C. W. Taylor and H. Busch. Physiol. Chem. Phys. 6:1 (1974).

30. R. K. Busch, I. Daskal, W. H. Spohn, M. Kellermayer and H. Busch.   Cancer Res. 34:2362 (1974).

31. J. Hilgers, R. C. Nowinski, G. Geering and W. Hardy.   Cancer Res. 32:98 (1972).

32. W. M. Bonner and R. A. Laskey.   Europ. J. Biochem. 46:83 (1974).

33. L. F. Cavalieri and E. Carroll.   Proc. Nat. Acad. Sci. USA 67:807 (1970).

34. J. A. Marmur.   J. Mol. Biol. 3:208 (1961).

35. T. O. Sitz, R. N. Nazar, W. H. Spohn and H. Busch. Cancer Res. 33:3312 (1973).

36. E. M. Wilson and T. C. Spelsberg. Fed. Proc. 32:608 (1973).

37. E. W. Johns and S. Forrester. Europ. J. Biochem. 8:547 (1969).

38. O. P. Samarina, E. M. Lukandin, J. Molnar and G. P. Georgiev. J. Mol. Biol. 33:251 (1968).

39. F. L. Yu and P. Feigelson. Proc. Nat. Acad. Sci. USA 69: 2833 (1972).

40. T. Pederson. J. Mol. Biol. 83:163 (1974).

41. J. J. Quinlan, P. Billings and T. E. Martin. Proc. Nat. Acad. Sci. USA 71:2632 (1974).

PROTEIN INTERACTION WITH DNA IN CHROMATIN

Barbara Sollner-Webb and Gary Felsenfeld
Laboratory of Molecular Biology
National Institute of Arthritis,
Metabolism and Digestive Diseases
Bethesda, Maryland 20014

Abstract

We have examined the arrangement of proteins on the DNA in chromatin isolated from duck reticulocytes and other tissues. The gross structure of chromatin, both purified and within nuclei, has been studied by digestion with staphylococcal nuclease. We found earlier that the DNA present in chromatin limit digests consists of a series of 12 fragments of very well defined size between 160 and 40 base pairs. We have now carried out digestion studies in nuclei, and confirm the observation of others that brief attack with nuclease results in appearance of large DNA fragments which are multiples of a subunit about 185 base pairs long. The kinetics of this digestion reveal that the higher multimers are precursors of the 185 base pair subunit, and that this subunit in turn degrades to produce a limit digest pattern very similar to that seen from chromatin. These results are consistent with a "beads on a string" model of chromatin, in which the "strings" are digested first, and the "beads" are then attacked to produce the characteristic limit digest pattern. We have made some progress toward interpreting this limit digest in terms of contributions from specific histones.

We have also examined the "fine structure" of chromatin protein organization in the neighborhood of an actively transcribed gene. This is done by isolating from duck reticulocyte chromatin both that fraction of the DNA resistant to nuclease, and the fraction titratable by polylysine. We find that a distinct segment of the globin gene is not accessible to polylysine; when the experiment is carried out with erythrocyte chromatin, no such region is noted. We have previously shown that transcription of duck reticulocyte chromatin in vitro leads to detectable globin RNA synthesis, while transcription of erythrocyte chromatin barely does. We believe that the polylysine titration results are directly related to the arrangement of proteins on "active" globin genes.

Chromatin is the complex of DNA and proteins isolated from eukaryotic nuclei. Its composition varies considerably with the biological source and the method of preparation. However, in the chromatin specimens we will be describing, the bulk of the protein consists of histones (about 1.25 grams/gram of DNA); only a small part (about 0.1 gram/gram DNA) is nonhistone protein. The histone components may be divided into the three familiar classes: lysine-rich histone H1 (largely replaced in reticulocytes and erythrocytes by another lysine-rich histone, H5); the slightly lysine-rich pair H2A/H2B; and the arginine-rich pair, H3/H4. Roughly speaking, about 40% of the histone mass is H2A/H2B, 40% is H3/H4, and 20% is H1 or H5. The most careful analyses presently available of histone content in chromatin from a variety of sources (1) suggest that H3 and H4 molecules may be present in equal numbers, but no simple stoichiometric relationship is detectable for the other histones.

The Gross Structure of Chromatin

Recent electron microscopic studies of chromatin (2) have suggested that much of the protein is distributed along the DNA as regularly spaced "beads" separated by string-like regions of DNA that are either protein-free or only lightly covered by protein. Van Holde and his collaborators (3,4) have shown that very mild digestion of chromatin with the enzyme staphylococcal nuclease results in the production of small nucleoprotein fragments which are almost certainly the repeating unit (bead + string) observed in intact chromatin. Other direct evidence for such a repeating structure has come from autolysis of nuclei (5) and treatment of nuclei with staphylococcal nuclease (6). Both procedures lead to production of a set of DNA fragments that are discrete in size and appear to be multiples of a fundamental subunit about 200 base pairs in length.

Studies of chromatin in our own laboratory began some years ago with an examination of the effects of staphylococcal nuclease on calf thymus chromatin (7). At that time we were interested in discovering what protection the histones and other nuclear proteins afforded the DNA against digestion by the nuclease. To our surprise we found that the attack proceeded quite reproducibly to an end point at which about half the DNA had been digested (Fig. 1). At this end point, the undigested DNA and all of the chromatin proteins are in the form of a precipitate; the proteins appear to be bound to the DNA, and there is evidence (7) that no protein exchange has occurred during the digestion process. The DNA in the limit digest is

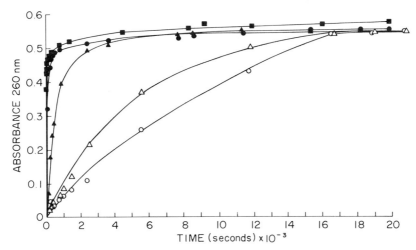

Fig. 1. Kinetics of digestion of calf thymus chromatin by staphylococcal nuclease. Release of perchloric acid soluble products as a function of time, using a variety of enzyme concentrations. Increasing rates of digestion correspond to 0.5, 1, 2, 10, and 100 μg/ml of enzyme. Solvent contained 5 mM sodium phosphate buffer (pH 6.7) and either 25 μM $CaCl_2$ (data at 100, 10, and 2 μg/ml of enzyme) or 11 μM $CaCl_2$ (data at 1 and 0.5 μg/ml of enzyme)(from 15).

about 110-150 base pairs (weight average) in length, and entirely double stranded.

What is the relationship of this limit digest to the larger repeating subunit structure revealed by mild nuclease digestion? In order to answer this question, we have carried out a study of the kinetics of nuclease digestion of nuclei and chromatin from very early times in the digestion process to the ultimate limit. We observe first (Fig. 2) that the kinetics of DNA digestion by staphylococcal nuclease are identical in nuclei and chromatin. (All of the following studies were carried out with nuclei and chromatin from duck reticulocytes). If nuclei are digested for a short time and the DNA is extracted and electrophoresed on 3% polyacrylamide gels, the pattern shown in Fig. 3 is obtained. This pattern is quite similar to the ones first reported by Williamson (8), Hewish & Burgoyne (5) and Noll (6). By calibration of the gels with H. influenza restriction enzyme digest fragments of phage λ and SV40 DNA of known length, and by sedimentation velocity measurements of the

215

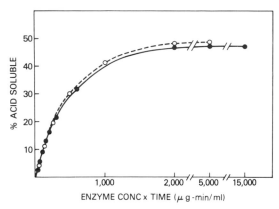

Fig. 2. Kinetics of nuclear digestion - release of acid soluble products. Duck reticulocyte nuclei (●) and duck reticulocyte chromatin (o) were digested with 1 to 60 µg/ml staphylococcal nuclease for 10 min to 1 hour in 1 mM Tris pH 8, 0.1 mM $CaCl_2$ at 1 mg DNA/ml. The fraction of the DNA soluble in 0.4 M perchloric acid, 0.4 M NaCl was measured. RNA was shown to contribute <3% of the A_{260} (from 10).

Fig. 3. Polyacrylamide gel electrophoresis of nuclear digest DNA. Nuclei were digested and the isolated DNA was run on 3% disc gels. From left to right are 1, 2, 3, 8, 16, 25, and 48% acid soluble digest. The last gel is a chromatin limit, 48%, digest (from 10).

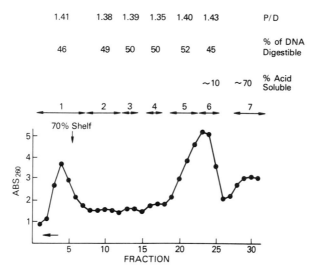

Fig. 4. Sucrose gradient centrifugation of partially digested nuclei. Nuclei were partially digested (7% acid soluble DNA), the reaction was stopped by the addition of Na EDTA to 0.3 mM, and the mixture was layered onto sucrose gradients. Direction of sedimentation is indicated by an arrow. The median sedimentation coefficient of the peak monomer fractions (S_{20}) was 10.2, measured in 1 mM Tris-HCl, 0.1 mM Na EDTA, pH 8 in the Model E ultracentrifuge (from 10).

DNA bands re-isolated from the gels, we find that the bands are multiples of a fundamental length of 185 base pairs, in good agreement with the results obtained by the above workers.

The nucleoprotein particles of the partial nuclear digest, as first reported by Noll (6), can be fractionated by sedimentation through sucrose gradients. Nucleoprotein fractions of increasing sedimentation coefficient (Fig. 4), when freed of protein and examined on polyacrylamide gels (Fig. 5) are found to contain successively monomer, dimer, trimer, and higher multimer DNA. It seems reasonable to suppose (as Van Holde and Noll have suggested) that the nucleoprotein fraction containing 185 base pair DNA is a bead-and-string monomer, that the nucleoprotein fraction containing 370 base pair DNA is the corresponding dimer, and so on. If the sucrose gradient frac-

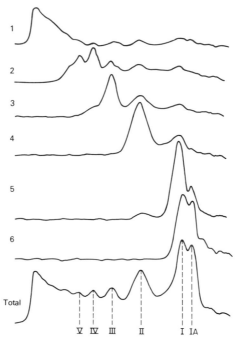

Fig. 5. Polyacrylamide gels of DNA from isolated nucleoprotein fragments:DNA isolated from the sucrose gradient fractions, pooled as indicated in Fig. 4, was run on 3% acrylamide
gels, stained, and photographed. Negatives were scanned with a
Joyce-Loebl microdensitometer. These scans have been shown
to be linear in DNA concentration. Migration is from left
to right. Scan numbers indicate the pooled fractions; "Total"
is DNA isolated from the material that was layered onto the
sucrose gradient (from 10).

tion containing pure nucleoprotein trimer is once again treated
with nuclease, it gives rise to dimer and monomer. Similarly,
purified dimer when digested gives rise to monomer. Almost
all of the chromatin must pass through the monomer during the
early stages of the digestion process. As indicated in Fig. 4,
the monomer (Band I) still has a protein:DNA ratio like that of
intact chromatin, and half of the DNA is still potentially digestible by nuclease.

Since this DNA is digestible, the monomer band is obviously
not the limit digest. If nuclease action is allowed to con-

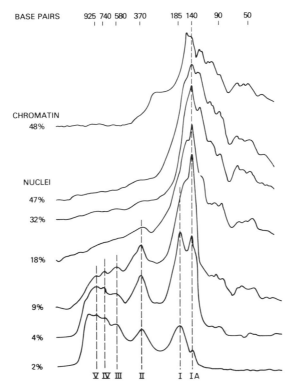

Fig. 6. Polyacrylamide gel electrophoresis of nuclear digest DNA. Nuclei were digested as in Fig. 2, isolated DNA was run on a 4% slab gel, stained, photographed and scanned. Migration is from left to right. From bottom to top are: nuclear digests at 2, 4, 9, 18, 32, and 47%, and a chromatin digest at 48% acid soluble DNA. Assigned sizes are shown across the top (from 10).

tinue, DNA band I (185 base pairs), is seen to degrade to a quite sharp band (IA) about 140 base pairs in length (Fig. 6). We believe that this reflects a process in which "strings" with an average length of 45 base pairs are preferentially degraded to produce "beads" that are free of strings, and that have DNA of quite uniform length. As the formation of band IA continues (Fig. 6) it becomes clear that the bead structure is susceptible to further degradation: a new set of discrete DNA bands begins to appear, ranging in size between about 140 and 40 base pairs. These bands become more intense as digestion pro-

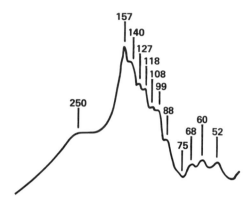

Fig. 7. Chromatin limit digest DNA. Chromatin was digested to the limit as in Fig. 2. The isolated DNA and sequenced
λ hin fragments were run on a 6% polyacrylamide gel. Nuclear
digest bands were visualized by staining; hin fragments by
autoradiography (from 10).

gresses, and eventually form the nuclear limit digest. The
entire pathway of nuclear digestion from early times to limit
digest is summarized in Fig. 6.

This limit digest band pattern of nuclei is almost identical to the limit digest pattern produced on the digestion of
isolated chromatin (9). A typical band pattern isolated from
a limit digest of duck reticulocyte chromatin is shown in Fig.
7. Differences have been observed between the digestion behavior of nuclei and chromatin: chromatin prepared by slow
swelling of nuclei followed by shearing does not pass through
the multimer-monomer steps shown in Fig. 6, and formation of
the limit digest bands begins at the earliest times of digestion (9). On the other hand, unsheared chromatin prepared by
osmotic shocking of nuclei behaves very much like intact nuclei, giving rise to patterns similar to those shown in the
early stages of digestion in Fig. 6 (10).

What is the origin of these DNA fragments of very well-
defined size present in the limit digest? We have begun to
examine this question by digesting DNA-protein complexes made
by reconstitution with selected histones or groups of histones.
If the entire complement of histones is used in reconstitution,

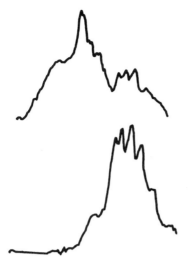

Fig. 8. Digestion of partial
reconstitute. Electrophoresis patterns of limit digest
of H3/H4 reconstituted with
DNA (lower). 0.52 g protein
per gram of DNA was present
in the starting complex. The
limit digest pattern of whole
chromatin is presented (above)
for comparison. (Manuscript
in preparation.)

subsequent digestion yields DNA patterns quite similar to that
of native chromatin digests, regardless of the source of the
DNA (9). If we now carry out reconstitution experiments using
only the arginine-rich histone pair H3/H4, the DNA of the limit
digest has a strikingly different appearance on polyacrylamide
gels (Fig. 8). Only four discrete DNA bands are observed.
These correspond to the smallest discrete fragments obtained
from total chromatin digests; the higher molecular weight group
of fragments (157-88 base pairs) is missing. Further reconstitution experiments are next performed using the same amounts
of H3/H4, and adding increasing quantities of the slightly
lysine rich histone pair H2A/H2B. The DNA gel patterns
resulting from digestion of these complexes are shown in Fig. 9.
As the amount of H2A/H2B is increased, some of the missing
higher molecular weight bands begin to appear, and the amount
of material in the lower molecular weight bands decreases. Our
preliminary experiments indicate that the H2A/H2B alone does
not give rise to bands in reconstitution experiments; we conclude that the higher molecular weight bands arise from interactions between the H3/H4 and H2A/H2B pairs. It should be
noted that the digest pattern from reconstitutes containing
equal weights of the two pairs (Fig. 9f) is still lacking some
of the bands obtained with normal chromatin, most notably the
highest molecular weight fragment 157 base pairs long. Addition of lysine rich histone (either H1 or H5) to the reconstitution mixture already containing H3/H4 and H2A/H2B results in
appearance of this missing band in limit digests. Lysine-rich

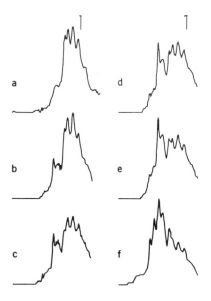

Fig. 9. Digestion of partial reconstitutes. Electrophoretic patterns of DNA from limit digests of reconstitutes. In each case 0.52 g of H3/H4 were present in the initial complex per gram of DNA. Patterns show the effect on the limit DNA digest of adding to this reconstitution mixture increasing amount of H2A/H2B. The initial ratios of total H2A/H2B to DNA (g/g) were: a) 0, b) 0.06, c) 0.11, d) 0.17, e) 0.24, f) 0.52. (Manuscript in preparation).

histone alone, when reconstituted with DNA, does not yield well-defined digest fragments.

Our results suggest that the pattern of the limit digest is generated by interaction of DNA with very well defined segments of histones, and that these segments are probably available only when the histone chains are bound to DNA in the proper conformation, an event which depends upon the interaction of the several histone species with one another. The complexity of the limit digest pattern of native chromatin (see also reference 11) suggests strongly that the monomer nucleoprotein fraction, although homogeneous by tests of protein:DNA ratio and sedimentation properties, may in fact contain more than one kind of particle. Within the restrictions of overall shape and protein content imposed upon the monomer, several combinations or arrangements of histones may be possible.

The Fine Structure of Chromatin.

We also wish to discuss briefly the arrangement of proteins in a special region of reticulocyte chromatin--the globin gene. It has been shown earlier in our laboatory and that of John Paul (12,13) that chromatin from duck reticulocytes or other hemoglobin-producing tissue can serve as a template in vitro for the production of globin messenger-like RNA, using E. coli

RNA polymerase. Erythrocyte chromatin, on the other hand, is nearly inactive as a template in globin message production (14).

In order to investigate the distribution of proteins in the neighborhood of the globin gene, we make use once again of nuclease digestion, isolating the limit digest of the chromatin. For convenience, we will call this "covered DNA". We can also isolate a second class of small DNA fragments from chromatin: those regions that react with polylysine. We have shown earlier (7,15) that just as half the DNA of chromatin is accessible to nuclease, half is also accessible to polylysine. If poly-D-lysine is used to titrate chromatin, the histones can be digested away afterward with Pronase (poly-D-lysine is resistant) and the exposed DNA is then removed with nuclease. The remaining DNA, freed of polylysine, we call "open" DNA.

What is the distribution of globin sequences in open and covered DNA from duck reticulocytes? This can be determined by carrying out annealing experiments in the presence of radioactively labelled DNA which is complementary in sequence to globin messenger RNA. This globin "probe" is prepared by the action of RNA-dependent DNA polymerase on globin messenger RNA and is identical to the probe used to detect the presence of globin-like RNA in in vitro RNA polymerase transcripts (12). When excess open or covered DNA from duck reticulocytes is denatured and annealed in the presence of globin probe, the results shown in Fig. 10 are obtained (14). There is a marked difference between the plateau values of the curves for open and covered DNA. This difference persists if still more denatured covered or open DNA are added to their respective annealing mixtures after the plateaus have been reached.

This result, together with appropriate controls (14), leads to the inescapable conclusion that a distinct portion of the globin gene in reticulocyte chromatin is inaccessible to polylysine, probably because it is itself covered by protein. In contrast, open and covered DNA isolated from erythrocyte chromatin show no differences in annealing characteristics, both reaching plateau values at about 80% of total probe present.

Are there any differences between the sequence populations of bulk open and closed DNA? To determine this, we radioactively label either open or covered DNA, and then anneal small amounts of the labelled product against large excesses of unlabelled open or covered DNA (Fig. 11). The annealing curves are identical to one another, indicating that sequences are common to the two kinds of DNA. Thus, the bulk of the

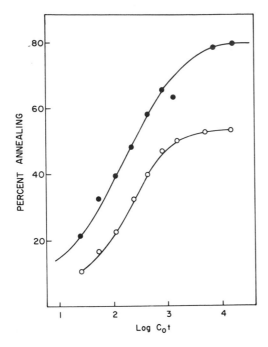

Fig. 10. Kinetics of annealing of globin cDNA to open and covered region DNA from duck reticulocyte chromatin: 0.1 ng of ^{3}H-globin DNA was reacted with 5 mg of either covered DNA ($\bullet$—$\bullet$) or open region (o—o) DNA from reticulocyte chromatin. Reactions were performed in 0.2 ml and duplex formation was assayed as described in legend to Figure 11 (from 14).

proteins of chromatin are distributed randomly with respect to base sequence, although as we discussed earlier, they may be spaced along the DNA fairly regularly relative to one another.

Fig. 11. Legend on next page.

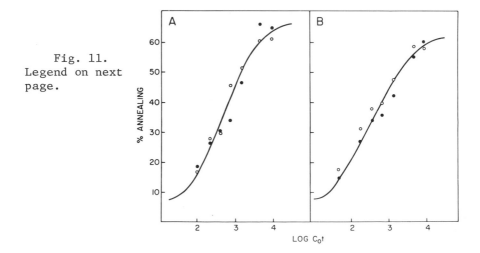

CONCLUSION

The bulk of chromatin structure may be thought of as involving subunits comprised of DNA and histone complexes, which although more or less regularly spaced along the DNA do not necessarily have a fixed relationship to a particular base sequence. We have examined this structure in some detail, and studied the mechanism of nuclease action starting with release of nucleoprotein oligomers, then monomers, and ending with the disruption of the monomer itself to produce the discrete fragments of the limit digest. These fragments appear to be generated by specific histone complexes; interaction among histones is necessary for their production.

In contrast, the arrangement of proteins in the neighborhood of the actively transcribed globin gene is clearly non-random: a specific segment of the gene is inaccessible to polylysine. In erythrocyte chromatin the difference disappears, and the globin gene behaves like the bulk of open and covered DNA. The arrangement of proteins in the neighborhood of actively transcribed genes is an important subject for future investigation.

References

1. S. Panyim, D. Bilek and R. Chalkley. J. Biol. Chem. 246: 4206 (1971).

2. A. Olins and D. Olins. Science 181:330 (1974).

3. R. Rill and K. Van Holde. J. Biol. Chem. 248:1080 (1973).

4. C. G. Sahasrabuddhe and K. E. Van Holde. J. Biol. Chem. 249:152 (1974).

5. D. Hewish and L. Burgoyne. Biochem. Biophys. Res. Commun. 52:504 (1973)

Fig. 11. Kinetics of annealing of open and covered DNA labelled in vitro with DNA polymerase III: Open and covered region DNA from reticulocyte chromatin were treated with exonuclease III and repaired with DNA polymerase III to yield DNA with a specific activity of 4×10^5 cpm/μg. A) ^{3}H-covered region DNA (0.02 μg) was reacted with 3 mg of either covered (●——●) or open (o——o) region reticulocyte DNA; B) ^{3}H-open region DNA (0.02 μg) was reacted with 3 mg of either covered (●——●) or open (o——o) region DNA. Annealing reactions were performed in a volume of 0.15 ml and duplex formation was monitored by single strand specific nuclease digestion (from 14).

6. M. Noll. Nature 251:249 (1974).

7. R. J. Clark and G. Felsenfeld. Nature New Biol. 229:101 (1971).

8. R. Williamson. J. Mol. Biol. 15:157 (1970).

9. R. Axel, W. Melchior, Jr., B. Sollner-Webb and G. Felsenfeld. Proc. Nat. Acad. Sci. USA 71:4101 (1974).

10. B. Sollner-Webb and G. Felsenfeld. Biochemistry, in press (1975).

11. H. Weintraub and E. Van Lente. Proc. Nat. Acad. Sci. USA 71:4249 (1974).

12. R. Axel, H. Cedar and G. Felsenfeld. Proc. Nat. Acad. Sci. USA 70:2029 (1973).

13. R. Gilmour and J. Paul. Proc. Nat. Acad. Sci. USA 70:3440 (1973).

14. R. Axel, H. Cedar and G. Felsenfeld. Biochemistry 14:2489 (1975).

15. R. J. Clark and G. Felsenfeld. Biochemistry 13:3622 (1974).

Note: Figures 1, 2, 3, 4, 5, 6, 7, 10 and 11 are reprinted with permission from Biochemistry. Copyright by the American Chemical Society.

CHROMATIN STRUCTURE AND TRANSCRIPTIONAL ACTIVITY

Joel Gottesfeld[+], Diane Kent*, Michael Ross* and James Bonner[+]
Divisions of Biology[+] and Chemistry and Chemical Engineering*
California Institute of Technology
Pasadena, California 91125

Abstract

Nuclease digestion studies with DNase II show that rat liver chromatin is organized into regions of DNA which differ in degree of susceptibility to enzymatic attack. The most sensitive segments of chromatin are enriched in sequences which code for RNA and thus are transcriptionally active. Nearly all the DNA in chromatin, however, is organized into regions of nuclease-sensitivity and resistance. The bulk of the DNA in chromatin is template-inactive; the nuclease resistant structures in transcriptionally inactive chromatin are nucleohistone complexes which resemble ν-bodies in the electron microscope. Nuclease-resistant structures in template-active chromatin are greatly enriched in nonhistone chromosomal proteins and RNA. These particles might correspond to ribonucleoprotein particles or transcription complexes. The distribution of nuclease-resistant complexes in template-active chromatin parallels that of inactive chromatin; however, thermal denaturation and circular dichroism studies suggest that active chromatin is in a more extended, DNA-like conformation than inactive chromatin.

Cellular differentiation in eukaryotes arises from the differential transcription and translation of genetic information. It is well established that transcription is restricted in chromatin as compared to DNA (1); moreover, recent evidence suggests that some of the mechanisms of genetic regulation remain intact in isolated chromatin (2,3). We believe that a comparison of the constituents and structure of transcriptionally active and inactive regions of chromatin will shed considerable light on mechanisms of gene activation and repression in eukaryotes. It is for this reason that we have developed methods for the isolation of the minor portion of chromatin which is transcriptionally active in vivo (4-6).

In this report we describe our procedure for chromatin fractionation and review findings on the sequence composition

227

of the template active fraction. It is the major purpose of this report, however, to relate transcriptional activity to chromatin structure. Using various biochemical and physical techniques we compare the structure of active and inactive regions of chromatin.

Chromatin Fractionation.

Previous studies from this laboratory have shown that the endonuclease DNase II preferentially attacks a minor fraction of chromatin DNA; the amount of DNA in this fraction is variable depending upon the source of the chromatin, but corresponds quite closely to the template activity of the particular chromatin as measured with exogenous RNA polymerase (5). We start our fractionation with sucrose-purified chromatin from rat tissues (7). The chromatin is dialyzed overnight against 25 mM Na acetate, pH 6.6. DNase II is added to the chromatin (10 enzyme units per A_{260} unit of chromatin) and digestion is allowed to proceed for various lengths of time. The reaction is stopped by raising the pH to 7.5 with 0.1 M Tris-Cl, pH 11. Unsheared chromatin is removed by centrifugation yielding a pellet (termed P1) and a supernatant (termed S1). The supernatant is further fractionated by the addition of $MgCl_2$ to 2 mM. The bulk of sheared chromatin is precipitated by either mono- or divalent cations (4,5) while active chromatin remains soluble. The inactive fraction is separated from active chromatin by centrifugation at 27,000 g for 10 min. The pellet (inactive chromatin) is termed P2 while the active chromatin supernatant is termed S2.

Figure 1 illustrates the time course of DNase II action on rat liver chromatin. At short times of enzyme treatment (inset, Figure 1), all the DNA liberated into fraction S1 fractionates into S2. After very short times (ca. 30 sec), 1.5% of the chromatin DNA is found in S2 and this DNA has a number-average length of 1800 base pairs. As the incubation time is increased, more of the DNA is found in fraction S2 and the DNA fragment length decreases. After a 5 min incubation, 11.3% of the total DNA is found in fraction S2; this DNA has a single-strand length of 400-500 base pairs. On longer times of digestion, P2 (Mg^{++} - insoluble material) is separated from the first pellet

---

Fig. 1. Time course of fractionation of rat liver chromatin. Chromatin was prepared and fractionated as described (6). The percent of total DNA in each chromatin fraction was determined by absorption at 260 nm of an aliquot diluted in 0.9 N NaOH. Pellet fractions were homogenized in 10 mM Tris-Cl, pH 8, prior to dilution for absorption determinations.

fraction (P1). On very long times, 80% of the chromatin DNA is sheared and found in fraction S1. The nature of the enzyme-resistant 20% is unknown. With rat liver chromatin, about 20% of the chromatin DNA is found in fraction S2 after a 15 min nuclease treatment; with rat ascites chromatin, 10% of the DNA is found in fraction S2. Thus, the percentage of total DNA in S2 is essentially the template activity of the particular chromatin sample as measured with exogenous RNA polymerase (5).

Sequence Composition of the Active Chromatin Fraction.

The reassociation kinetics of DNA from unfractionated chromatin and fraction S2 (5-min nuclease exposure) have been determined. Both DNAs have rapidly, intermediately and slowly renaturing kinetic components, representing about 10%, 20%, and 70% of the input DNA, respectively. In the DNA of unfractionated chromatin these components correspond to highly repetitive, moderately repetitive and single copy sequences. The intermediate and slow components of fraction S2 DNA renature at rates 4 to 7-fold faster than their respective counterparts in unfractionated chromatin DNA. The C_0t 1/2 values for the intermediate component of whole rat DNA and fraction S2 DNA are 1.05 and 0.26, respectively. The C_0t 1/2 values for the slow component of whole rat DNA and S2 DNA are 1500 and 225, respec-

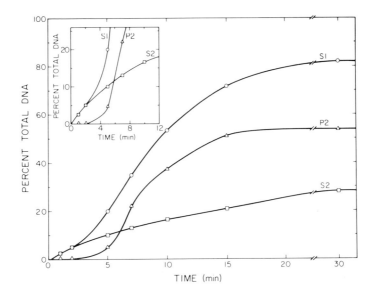

Fig. 1.   Caption on previous page.

tively. These differences in renaturation rates are not due to differences in reaction conditions or DNA size. A random population of DNA sequences would renature with the same kinetics as whole rat DNA; therefore, we interpret the observed rates of renaturation to mean that fraction S2 contains a limited fraction of the complexity of the rat genome--that is, a specific subset of DNA sequences.

We have isolated the slow component of S2 DNA, labelled this DNA in vitro with ^{125}I, and determined the reassociation kinetics of this tracer in the presence of an excess of unfractionated rat DNA. The S2 slow component tracer reanneals with the single copy sequences of whole DNA. This shows that the S2 slow component is derived from single copy sequences and not repetitive sequences. We calculate that the S2 slow component contains about 10-15% of the single copy complexity of rat DNA. In a separate report, we show that the repetitive sequences of fraction S2 comprise about one-tenth of the repetitive complexity of whole rat DNA (8). Preliminary data suggest that the repetitive and single copy sequences of fraction S2 DNA are interspersed on a fine scale. About 70% of 1800-nucleotide long S2 fragments contain repetitive sequences, while only 30% of 400-nucleotide long S2 fragments contain repeated DNA (8).

Cross-hybridization experiments with the DNAs of the chromatin fractions have shown that fraction S2 contains only about 30% of its sequences in common with fraction P1 (5-min nuclease treatment). Furthermore, those P1 DNA sequences which are present in S2 DNA are found in low abundance in fraction S2. Thus, fractionation according to DNA sequence has been accomplished by the methods employed. To determine to what degree the DNA sequences of fraction S2 are tissue specific, we have isolated this fraction from both liver and brain chromatin. Cross-hybridization experiments show that about 30-40% of the sequences of brain fraction S2 are also present in the liver fraction; however, these brain S2 sequences are found less frequently in liver S2 DNA than in brain S2 DNA. Thus, fractionation is also tissue-specific.

RNA-DNA Hybridization Experiments

We have investigated the degree of homology of single copy DNA isolated from the liver chromatin fractions with RNA from both liver and brain. The liver S2 chromatin fraction is enriched 4 to 6-fold over fraction P1 in single copy sequences which code for liver total cellular RNA. On the other hand, the liver fraction S2 is not enriched in sequences which code for brain cellular RNA (8).

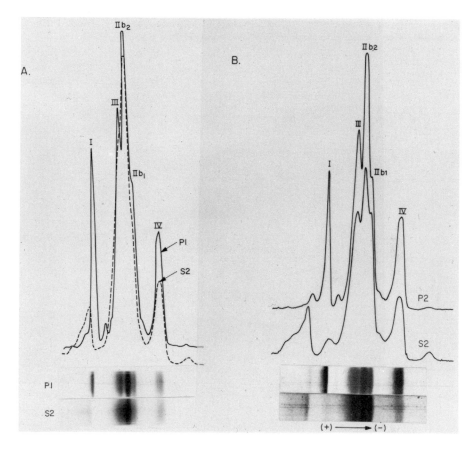

Fig. 2. Electrophoretic profiles of acid soluble protein from active and inactive fractions of chromatin. Urea-disc gel electrophoresis was performed according to Panyim and Chalkley (32). A. Fractions S2 and P1 obtained after 5-min nuclease treatment. B. Fractions S2 and P2 obtained after 15-min nuclease treatment.

In summary, we have found that fraction S2 is composed of a subset of single copy DNA sequences and a subset of families of repeated DNA sequences. From RNA-DNA hybridization experiments, nearly 60% of the single copy sequences of fraction S2 appear to be transcribed in vivo. We conclude from these data that the distribution of repeated DNA sequences with respect to transcribed single copy sequences is nonrandom. The significance of this finding with respect to models of gene regulation (9) is discussed elsewhere (8).

Chemical Composition of the Chromatin Fractions.

It has been reported previously that the active fraction is enriched in nonhistone chromosomal proteins and depleted in histone protein as compared to unfractionated chromatin (4,6). The histone complement of fraction S2 is about half that of unfractionated chromatin. Histone I is totally absent from fraction S2 and, in some determinations, histone IV is found in slightly diminished amounts (Figure 2). A protein band which migrates slightly slower than histone I is seen in the gels of S2 acid-soluble protein. The nature of this band is not known; however, Garrard and Bonner (10) have noted that a band at a similar position in whole rat liver histone extracts turns over at a rate 10-fold greater than histone I.

There is a two-to-three-fold enrichment in nonhistone protein in fraction S2 over whole chromatin (6). Several qualitative and quantitative differences can be seen in the SDS-polyacrylamide gels of the nonhistone proteins of active and inactive chromatin (Figure 3). The most striking difference is the enrichment in fraction S2 of two bands in the molecular weight range of 38-40,000. These polypeptides co-electrophorese with two of the proteins involved in the packaging of heterogeneous nuclear RNA (11). Several bands in the 90-100,000 molecular weight range are also enriched in fraction S2. The inactive fractions P1 and P2 appear to be enriched in polypeptides of molecular weight 28,000 and 65,000. The former polypeptide has been isolated from whole chromatin and characterized as a highly acidic protein; the latter has also been isolated in pure form and tentatively identified as the protein actin (Douvas, Herrington and Bonner, in preparation). A band in P1 chromatin (Figure 3A; 5-min nuclease digestion) at 30-32,000 molecular weight appears to be greatly diminished in P2 chromatin (Figure 3B; 15-min nuclease treatment); perhaps this band is specific to P1 chromatin. Comparisons in the 45-55,000 molecular weight range are difficult due to the large amount of protein found in this region.

Subunit Structure of Chromatin.

The idea of a regular repeating structure in chromatin is suggested from the X-ray diffraction studies of Pardon, Wilkins and Richards (12). A series of reflections are seen at 110, 55, 37, 27 and 22 Å in the X-ray patterns of native and reconstituted nucleohistones, but not in the X-ray diffraction patterns of DNA or histone by themselves. It was proposed that chromatin DNA was organized into a regular supercoil of pitch

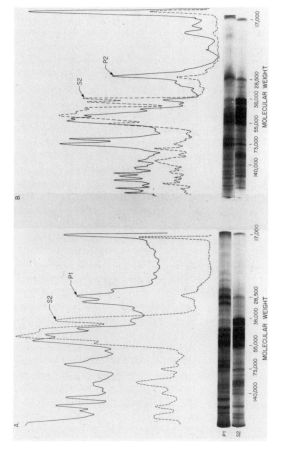

Fig. 3. SDS-polyacrylamide gel electrophoretic profiles of acid-insoluble protein from active and inactive chromatin fractions. Disc gel electrophoresis was performed according to King and Laemmli (33). A. Fractions S2 and P1 obtained after 5-min nuclease treatment. B. Fractions S2 and P2 obtained after 15-min nuclease treatment.

100-120 Å and diameter 100 Å. This model, although widely accepted for some time, has recently come under question. Olins and Olins (13) have observed regular spacings of chromatin particles (termed ν-bodies) in water-swollen nuclei centrifuged onto electron microscope grids. These particles are 60-80 Å in diameter and are joined by thin "filaments" 15 Å in diameter. Nuclease digestion studies also support a subunit or particulate structure for chromatin. Hewish and Burgoyne (14) and Burgoyne et al. (15) have found that an endogenous Ca^{++}-Mg^{++}-activated endonuclease of rat and mouse liver nuclei cleaves chromatin DNA at regular intervals; fragments of DNA at integral multiples of about 200 base pairs were observed. Many workers have studied the digestion of isolated chromatin with exogenous nucleases; Clark and Felsenfeld (16,17) have reported that staphylococcal nuclease digests 50% of chromatin DNA regardless of the source of the chromatin. The nuclease-resistant fragments were found to contain double-stranded DNA, 110-120 base pairs in length. These results have been confirmed and extended by van Holde and his co-workers (18-20). Moreover, isolated particles from nuclease-treated or sonicated chromatin resemble ν-bodies in the electron microscope (20,21).

Recent evidence obtained from neutron scattering of chromatin (22) suggests that the 110 Å reflection seen in the X-ray pattern does not arise from a DNA repeat, but rather from a regular spacing of a protein core. Baldwin et al. (22) and van Holde et al. (23) propose that chromatin DNA is wound about the exterior of the protein core. We are interested in whether the template-active fraction of chromatin is organized in a structure similar to the bulk of chromatin or whether it is in a different structure.

To consider the structure of the active regions of chromatin, we must first examine the DNase II digestion products of unfractionated chromatin. The bulk of chromatin DNA in any given cell type is transcriptionally inactive; therefore, the properties of unfractionated chromatin reflect the structure of inactive regions. We have digested rat liver chromatin with DNase II for extended periods of time (60-90 min), removed unsheared chromatin by low speed centrifugation, and analyzed the sheared chromatin on isokinetic sucrose gradients (Figure 4). About 40% of the sheared chromatin DNA sediments extremely slowly; the bulk of this DNA is acid soluble and hence has been reduced to oligonucleotides by the nuclease. The remaining chromatin appears to sediment at about 12-13S, with some material sedimenting slightly faster. We have studied the properties of nuclease-resistant material from both native and histone I-depleted chromatin (Table I). The nuclease-resistant species from histone I-depleted chromatin appears to sediment

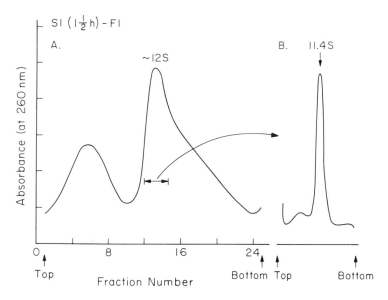

Fig. 4.   Sucrose sedimentation of DNase II-sheared chromatin.   A.   Chromatin from rat liver was prepared by the method
of Marushige and Bonner (7).   The chromatin was Virtis-sheared
and histone I was extracted with 0.5 M NaCl (see Table I).   The
resultant chromatin was dialyzed against 25 mM Na acetate, pH
6.6; DNase II was added to 10 units per A_{260} unit of chromatin.
Digestion was terminated after 60-90 min by raising the pH to
7.5 with 0.1 M Tris-Cl, pH 11.   Unsheared chromatin (about 20%
of the input DNA) was removed by centrifugation at 27,000 g
for 10 min.   The supernatant was layered on isokinetic sucrose
gradients in SW 25.1 cellulose nitrate tubes.   The gradients
were formed according to Noll (24); the parameters were C_{TOP} =
15% (w/v), C_{RES} = 34.2% (w/v), V_{MIX} = 31.4 ml.   All solutions
contained 10 mM Tris-Cl, pH 8.   Centrifugation was at 24,000
rpm for 42 h.   B.   The "monomer" fractions from the first gradient were pooled, dialyzed against 10 mM Tris, concentrated
with an Amicon device, and layered on a second isokinetic sucrose gradient.   The parameters for the second gradient were
C_{TOP} = 5.1% (w/v), C_{RES} = 31.4% (w/v), and V_{MIX} = 9.4 ml for
the SW 41 rotor.   Centrifugation was at 39,000 rpm for 16.5 h
at 4°.   The absorbance profile shown was obtained with an Isco
Ultraviolet Analyzer and chart recorder.

235

TABLE I

Properties of Isolated Chromatin Particles[a]

Measured Property	Chromatin	
	+ Histone I	Histone I-Depleted[b]
Observed Sedimentation Values	13.3 ± 0.4S (3)[c]	12.2 ± 0.6S (14)
Particle Size		
Normal	80 ± 17 Å (125)	87 ± 19 Å (100)
Formaldehyde-Fixed	96 ± 23 Å (50)	81 ± 13 Å (85)
DNA Size (Nucleotides)[d]	211 ± 9 (3)	166 ± 16 (8)
ρ_{CsCl} (gm/cm^3)[e]	1.423	1.440 ± 0.014 (3)
Protein to DNA Ratio (w/w)		
From ρ_{CsCl}[f]	1.13	0.96
Chemical Estimate[g]	0.9 - 1.0 (89-95% histones)	0.92
T_m (0.2 mM EDTA, pH 8)	-	78°C (T_mDNA = 42°)
$\Delta\varepsilon$275 nm	-	0.9 - 1.0 ($\Delta\varepsilon$ DNA = 2.6)

[a]Chromatin was treated with DNase II for 60-90 min and then layered on isokinetic sucrose gradients (Figure 4). [b]For the removal of histone I, Virtis-sheared chromatin (45v, 90 sec) was extracted with 0.5 M NaCl. The resultant nucleohistone was pelleted by centrifugation in the Ti50 rotor at 50,000 rpm for 18 h. The pellet was homogenized in 10 mM Tris-Cl, pH 8, and treated in the same manner as whole chromatin. [c]Mean ± standard deviation; number of trials in parentheses. (Caption continued at bottom of next page.)

236

slightly slower than that from native chromatin. When the nuclease-resistant peak from histone I-depleted chromatin was
rerun on 5-20% isokinetic sucrose gradients (Figure 4B), a
sedimentation value of 11.5 ± 0.3 S was observed. The gradients
were calculated for particles of density 1.4 gm/cm^3 (24). The
observed sedimentation value is in close agreement with the
values reported by other workers for the nuclease-resistant
particles obtained with the micrococcal enzyme (19,20,25).

The nuclease-resistant fractions obtained from sucrose
gradients (Figure 4) have been further characterized by electron microscopy (Figure 5). Although roughly circular in outline, the particles show little detail. The particles from
native and histone I-depleted chromatin appear identical in the
electron microscope. Further, fixation of the nuclease-resistant particles with formaldehyde (26) does not affect their
appearance. We conclude that there is no statistical difference between the diameters of the particles from native or
histone I-depleted chromatin; furthermore, formaldehyde fixation does not alter the observed diameters (Table I). These
diameters are close to the values reported for ν-bodies in
unsheared chromatin (13) and for the particles obtained from
either sonicated chromatin (21) or staphylococcal nuclease-
treated chromatin (20). Thus both DNase II and the micrococcal
enzyme liberate particles from chromatin which resemble the
ν-bodies of Olins and Olins (13).

Table I lists some additional properties of the DNase II-
resistant particles. From the buoyant density of formaldehyde-
fixed particles, we calculate protein:DNA ratios of 0.96 to
1.13. These values agree within about 10% with the ratios determined chemically on unfixed material; furthermore, 89-95%
of the protein in the nuclease-resistant particles is acid-

Table I footnotes continued. [d]Single-strand lengths were
determined by akaline sedimentation velocity in isokinetic sucrose gradients (24). The parameters used were C_{TOP} = 15.9%
(w/v); C_{RES} = 38.9% (w/v); V_{MIX} = 6.1 ml. Samples (100-200 µl)
in 0.1 N NaOH were layered on gradients and centrifuged in the
SW 50.1 rotor at 48,000 rpm for 16 h at 20°. Lengths were determined relative to a standard sized by electron microscopy
(a gift of Ms. M. Chamberlin). [e]CsCl-density gradients were
formed by centrifugation in the SW 50.1 rotor at 46,000 rpm for
60 h at 20°. The initial density of the CsCl was 1.44 gm/cm^3.
Samples were fixed with formaldehyde (26) prior to centrifugation. [f]Protein to DNA ratio calculated from density by the
method of Brutlag et al. (26). [g]Chemical composition was performed by the methods of Bonner et al. (1).

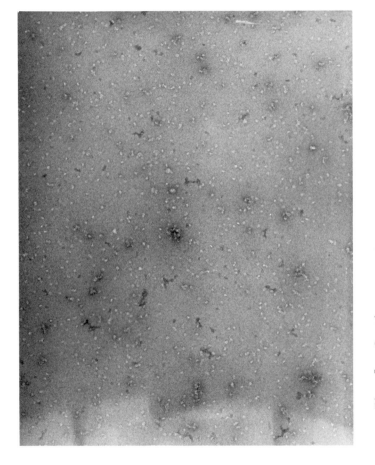

Fig. 5. Caption on next page.

extractable histone protein. The acid extracts of particles from histone I-depleted chromatin contain the slightly-lysine and arginine-rich histones in about the same proportions as in whole chromatin. We have not, as yet, determined whether the particles from native chromatin contain histone I. From the lengths of DNA contained in the nuclease-resistant particles and protein:DNA ratios (Table I), we calculate 295,000 and 210,000 for the molecular weights of particles from native and histone I-depleted chromatin, respectively. This calculation assumes that the DNA in the resistant particles is double stranded (single-strand lengths were determined); this is reasonable, since the particles exhibit native hyperchromicity on thermal denaturation. The length of DNA in particles from native chromatin (211 ± 9 base pairs) is consistent with the findings of Noll (25) and Senior et al. (21). Further, Noll has found a second nuclease-sensitive site within chromatin subunits; nuclease attack at this internal site yields a DNA fragment 170 base pairs in length, nearly the size of the DNA isolated from the particles of histone I-depleted chromatin (Table I). Thus histone I appears to protect 30-40 base pairs of DNA from nucleolytic attack. In contrast to these results, Clark and Felsenfeld (16,17) and Sahasrabuddhe and van Holde (19) report a DNA length of 110-120 base pairs for staphylococcal nuclease-resistant fragments of chromatin.

Figure 6 illustrates densitometer scans of stained polyacrylamide gels of DNA isolated after extensive digestion of chromatin with DNase II. Nuclease digestions were at twice the usual enzyme concentration and incubations were for 90 min. Resistant particles from native and histone I-depleted chromatin were studied. In the native chromatin digest (upper scan, Figure 6), bands at 174, 124, 75 and 46 nucleotide pairs can be seen. A shoulder at 200-250 base pairs is also seen. In contrast, the band at 174 base pairs and the shoulder at 200-250 base pairs are greatly diminished in the DNA gel of histone I-depleted chromatin particles. A band at 96 base pairs, not seen in native chromatin digest, is observed in the histone I-depleted chromatin digest. Low molecular weight material (25-50 base pairs) is seen in both gels. Axel et al. (27) have

---

Fig. 5. Electron micrograph of "monomer" chromatin particles. Chromatin was sheared with DNase II and fractionated on sucrose gradients as described (Figure 4). The "monomer" fraction was fixed with formaldehyde (26) and examined by electron microscopy by the methods of van Holde et al (20). The sample was first stained with uranyl oxalate (1%, pH 7), then poststained with uranyl acetate (1%, pH 4.35).

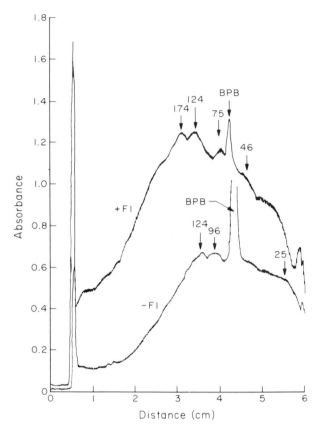

Fig. 6.   Densitometer scans of polyacrylamide gels of DNA isolated from monomer chromatin subunits.   Chromatin was digested extensively (90 min at 20 enzyme units per A_{260} unit of chromatin) and centrifuged in sucrose gradients as described (Figure 4).   DNA was prepared from the monomer fractions by banding in CsCl gradients.   The initial CsCl concentration was 1.65 gm/cm^3 and the gradients were formed by centrifugation at 38,000 rpm in the SW rotor at 22°C for 60 h.   The DNA-containing fractions were pooled, dialyzed extensively to remove CsCl, and lyophilized.   The DNA was dissolved in a small volume and electrophoresed on 4% acrylamide gels containing 0.2% bis-acrylamide.   The buffer was 0.02 M Na acetate, 0.04 M Tris-acetate, 2 mM EDTA, 1 mM Mg acetate, pH 8.   Electrophoresis was carried out at 2 mA/gel at 2°C.   The gels were stained in ethidium bromide (0.1 µg/ml) for 2 h, photographed by transmission with a short wavelength UV lamp, and the photographic negatives were

presented gels of chromatin DNA after digestion with staphylococcal nuclease. Bands at 200, 128, 76 and 48 base pairs are
observed. Removal of lysine-rich histones from chromatin results in the disappearance of the bands at 200 and 128 base
pairs. Similar to our finding, the major bands present after
removal of lysine-rich histones are in the range of 80-120 base
pairs. It must be concluded that these DNA bands arise from
real components of chromatin structure since two enzymes give
qualitatively similar results; staphylococcal nuclease, however, does not appear to recognize the 174 base pair structure.

The DNA in the nuclease-resistant particles is highly
stablized against thermal denaturation; melting transitions at
76° and 79° can be recognized in the derivative plot of a melting profile of particles from histone I-depleted chromatin.
The overall T_m is 78°, nearly 40° above the T_m of deproteinized
DNA in the same solvent. The nuclease-resistant particles
exhibit a nonconservative circular dichroism spectrum, suggesting that the DNA might be in a conformation different from that
of the B-form in aqueous solution. Protein-induced folding of
B-form DNA could also account for the observed CD spectrum.

From the calculated molecular weights, the nuclease-resistant particles from histone I-depleted chromatin could accommodate two each of the slightly-lysine and arginine-rich histones
(total molecular weight ≈ 108,760). The particles from native
chromatin could accommodate these histones plus histone I (130,
400 total molecular weight). Kornberg (28) has proposed that
chromatin subunits contain two each of the slightly-lysine-rich
histones (IIb1 and IIb2), two each of the arginine-rich histones (III and IV), perhaps one molecule of histone I and 200
base pairs of DNA. Our findings and those of others (21) are
generally consistent with this model; however, we do not find
equimolar amounts of the arginine and slightly-lysine-rich histones in either whole rat liver chromatin or isolated ν-bodies
as predicted by Kornberg (28). Further work on the distribution of histones in the subunits needs to be carried out.

Structure of Transcribed Chromatin.

We now turn attention to the structure of the transcribed
regions of chromatin. It was reported previously (29) that the
active fraction S2 exhibits a circular dichroism spectrum more
like that of B-form DNA than either unfractionated chromatin or

---

scanned. DNA sizes were determined relative to HIN I restriction fragments of φX174 DNA. The tracking dye was bromphenol
blue (BPB).

TABLE II

Properties of Active Chromatin Sub-Fractions[a]

Fraction	Fraction of Total S2 DNA	Observed Sedimentation Value	Average Melting Temperature (°C)[b]	Chemical Composition Relative to DNA (w/w)[c]				
				DNA	RNA	Total Protein	Acid Soluble Protein	Acid Insoluble Protein
Total S2 (5 min)	100%	-	54.5	1.0[e]	0.25	2.21	0.61	1.60
Light	53%	-	45.6	1.0	-	0.88 ± 0.06	0.24	0.60
Intermediate	} 47%	15.1 ± 0.7S (8)[d]	51	1.0	0.4		0.72	1.35
Heavy		20.9 ± 0.7S (8)	56	1.0	0.7	2.94 ± 0.06	0.54	3.2

[a]Chromatin from rat liver (7) was treated with DNase II for 5 min and fractionated as described (6). Fraction S2 was layered on isokinetic sucrose gradients as described in the legend to Figure 7. [b]The average melting temperature was measured in 0.2 mM EDTA, pH 8. [d]Mean ± standard deviation; number of trials in parentheses. [e]Data of Gottesfeld et al. (6). For total S2 chromatin, all acid soluble protein indicated is histone; however, we do not know what fraction of the acid soluble protein of the subfractions is histone protein.

242

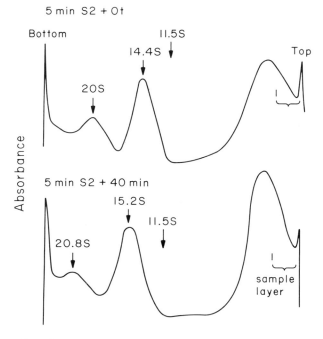

Fig. 7. Sucrose gradient sedimentation of the template-active fraction of rat liver chromatin. Chromatin was isolated and fractionated into inactive and active regions (6). The active fraction S2 was concentrated on an Amicon device, layered on an isokinetic sucrose gradient (Figure 4B), and centrifuged in the SW 41 rotor at 39,000 rpm for 17.5 h at 4°. Sedimentation is from right (TOP) to left (BOTTOM). The absorbance profiles shown were obtained with the Isco UV analyzer and chart recorder. In the lower scan, DNase II was reactivated after the isolation of fraction S2. Reactivation was accomplished by adding EDTA to 20 mM and lowering the pH to 6.4. Incubation was for an additional 40 min at 24°. The reaction was terminated by raising the pH to 8 with Tris. The initial DNase treatment used for chromatin fractionation was for 5 min at 10 units per A_{260} unit of chromatin.

the inactive fractions (P1 and P2). Similar results have been presented by Polacow and Simpson (30) for fractions obtained by ECTHAM-cellulose chromatography of sonicated chromatin. Further, we have found that fraction S2 melts at temperatures only slightly higher than DNA (Table II); a melting transition at 55° in whole chromatin has been assigned to DNA complexed with the nonhistone chromosomal proteins (31).

We now ask whether any nuclease-resistant particulate
structures exist in fraction S2. Chromatin from rat liver was
treated with DNase II for 5 min, fractionated as before, and
S2 material analyzed on isokinetic sucrose gradients (Figure 7).
The average single-strand length of S2 DNA after 5 min of nuclease treatment is 400-500 nucleotides; however, about half of
the UV-absorbing material applied to the gradient does not
sediment appreciably. This material, however, is all acid pre-
cipitable. Two rapidly sedimenting peaks are seen in the gradient of S2 chromatin: one at 14-15S and another 20-21S.
About 6% of the input nucleic acid has pelleted during this
centrifugation. DNase treatment of chromatin is required for
the appearance of the 14-21S particles.

In order to determine whether fraction S2 contains 11-12S
"ν-body" particles, we have redigested S2 chromatin with DNase
II. This was accomplished by reactivating the DNase in fraction S2 by the addition of EDTA (pH 6.4). Reincubation of S2
chromatin was carried out at 24° for 20-120 min. Digestion was
terminated by raising the pH to 8.0. The redigested S2 chromatin was then analyzed on isokinetic sucrose gradients (Figure 7). No significant changes were observed in the >10S
region of the gradient. Table II shows that the sedimentation
values observed in the two gradients of Figure 7 are within
the range of S values seen in a total of 8 similar gradients.

We have determined the chemical composition of the subfractions of S2 chromatin (Table II). The material at 14-21S
is enriched in both RNA and nonhistone chromosomal proteins.
Some acid-soluble protein is present in the 14-21S fractions.
We are currently investigating the nonhistone chromosomal proteins of these fractions. Preliminary data suggest that the
14-21S particles might correspond to the ribonuclear protein
particles (RNP) described by Martin et al. (11). It will be
recalled that S2 chromatin is enriched in polypeptides which
co-electrophorese with the RNA packaging proteins. The presence of DNA and RNA in these particles could indicate that we
have isolated transcription complexes; alternatively, the DNA
could be bound to the RNPs nonspecifically. This is a matter
for further investigation.

On redigestion of S2 chromatin, it was found that 50-60%
of the UV-absorbing material became acid soluble (Figure 8).
Redigestion has been performed in three ways: reactivation of
DNase II; addition of new DNase under normal conditions of
enzyme treatment (pH 6.6); and addition of staphylococcal nuclease. In all cases, 50-60% of the input S2 chromatin A_{260}
became acid soluble. The remaining nuclease-resistant chroma-

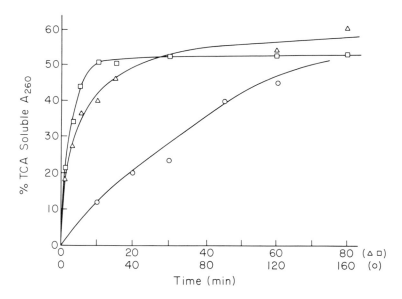

Fig. 8.   Kinetics of digestion of chromatin S2.   Chromatin from rat liver was prepared and fractionated into active (S2) and inactive segments following treatment with DNase II for 5 min (6).   DNase II was reactivated as described in the legend to Figure 7 (o).   Aliquots of the active fraction S2 were dia-lyzed against either 25 mM Na acetate, pH 6.6, or 5 mM Na phos-phate, pH 6.7, plus 2.5 x 10^{-4} M MgCl .   DNase II ($\triangle$) was added to the Na acetate-chromatin at 10 units per A_{260} unit of S2 chromatin; Staphylococcal nuclease ($\square$) was added to the Na phosphate-chromatin at 50 units per ml.   Reactions were carried out at 24°.   Aliquots were taken at various times to test for the production of trichloroacetic acid-soluble material (measured by absorbance at 260 nm).

tin DNA was found to have a weight-average single strand length of 120 nucleotides (DNase II or staphylococcal nuclease). Thus, it is likely that the template active fraction is organized in a fashion which parallels that of inactive chromatin DNA; that is, the active fraction is composed of regions of nuclease-sensitive and nuclease-resistant DNA. Nuclease-resistant segments in transcriptionally inactive chromatin are due to histone-DNA interactions, while the resistant portions of ac-tive chromatin must arise from the interaction of DNA with non-histone chromosomal proteins.

If we define ν-bodies as histone-rich, RNA-free, nucleo-protein complexes which sediment at 11-13S, we conclude that no such structures exist in the template-active fraction of chromatin.

Acknowledgements

We thank Dr. W. T. Garrard for help with the gel electrophoresis, and Mr. William Pearson for a critical evaluation of this manuscript. This work was supported by the U.S. Public Health Service (grants GM 86, GM 19984, GM 13762 and GM 01262).

References

1. J. Bonner, R. Chalkley, M. Dahmus, D. Fambrough, F. Fujimura, R. C. Huang, J. Huberman, R. Jensen, K. Marushige, H. Ohlenbusch, B. Olivera and J. Widholm. In: "Methods in Enzymology", Vol. XII Part B, L. Grossman and K. Moldave, eds. Academic Press, New York (1968), p. 3.

2. R. Axel, H. Cedar and G. Felsenfeld. Proc. Nat. Acad. Sci. USA 70:2029 (1973).

3. R. S. Gilmour and J. Paul. Proc. Nat. Acad. Sci. USA 70: 3440 (1973).

4. K. Marushige and J. Bonner. Proc. Nat. Acad. Sci. USA 68:2941 (1971).

5. R. J. Billing and J. Bonner. Biochim. Biophys. Acta 281: 453 (1972).

6. J. M. Gottesfeld, W. T. Garrard, G. Bagi, R. F. Wilson and J. Bonner. Proc. Nat. Acad. Sci. USA 71:2193 (1974).

7. K. Marushige and J. Bonner. J. Mol. Biol. 15:160 (1966).

8. J. M. Gottesfeld, G. Bagi, B. Berg and J. Bonner. Submitted for publication (1975).

9. E. H. Davidson and R. J. Britten. Quart. Rev. Biol. 48: 565 (1973).

10. W. T. Garrard and J. Bonner. J. Biol. Chem. 249:5570 (1974).

11. T. Martin, P. Billings, A. Levey, S. Ozarslan, T. Quinlan, H. Swift and L. Urbas. Cold Spring Harbor Symp. 38:921 (1973).

12. J. F. Pardon, M. H. F. Wilkens and B. M. Richards. Nature 215:508 (1967).

13. A. L. Olins and D. E. Olins. Science 183:330 (1974).

14. D. R. Hewish and L. A. Burgoyne. Biochem. Biophys. Res. Commun. 52:504 (1973).

15. L. A. Burgoyne, D. R. Hewish and J. Mobbs. Biochem. J. 143:67 (1974).

16. R. Clark and G. Felsenfeld. Nature 229:101 (1971).

17. R. Clark and G. Felsenfeld. Biochemistry 13:3622 (1974).

18. R. Rill and K. E. van Holde. J. Biol. Chem. 248:1080 (1973).

19. C. G. Sahasrabuddhe and K. E. van Holde. J. Biol. Chem. 249:152 (1974).

20. K. D. van Holde, C. G. Sahasrabuddhe, B. Shaw, E. F. J. van Brugger and A. C. Arnberg. Biochem. Biophys. Res. Commun. 60:1365 (1974).

21. M. B. Senior, A. L. Olins and D. E. Olins. Science 187: 173 (1975).

22. J. P. Baldwin, P. G. Boseley, E. M. Bradbury and K. Ibel. Nature 253:245 (1975).

23. K. E. van Holde, C. G. Sahasrabuddhe, B. R. Shaw. Biochem. Biophys. Res. Commun., in press (1975).

24. H. Noll. Nature 215:360 (1967).

25. M. Noll. Nature 251:249 (1974).

26. D. Brutlag, C. Schlehuber and J. Bonner. Biochemistry 8:3214 (1969).

27. R. Axel, W. Melchior, Jr., B. Sollner-Webb and G. Felsen-feld. Proc. Nat. Acad. Sci. USA 71:4101 (1974).

28. R. Kornberg. Science 184:868 (1974).

29. J. M. Gottesfeld, J. Bonner, G. Radda and I. O. Walker. Biochemistry 13:2937 (1974).

30. I. Polacow and R. T. Simpson. Biochem. Biophys. Res. Commun. 52:202 (1973).

31. Hsueh-Jei Li and J. Bonner. Biochemistry 10:1461 (1971).

32. S. Panyim and R. Chalkley. Arch. Biochem. Biophys. 130: 337 (1969).

33. J. King and U. K. Laemmli. J. Mol. Biol. 62:465 (1971)

INTERACTIONS OF A SUBCLASS OF NONHISTONE CHROMATIN PROTEINS WITH DNA

Gordhan L. Patel and Terry L. Thomas
Department of Zoology
University of Georgia
Athens, Georgia   30602

Abstract

The _in vitro_ DNA binding parameters of a subclass (APNH) of nonhistone chromatin proteins with high affinity for DNA have been examined by the nitrocellulose membrane filtration method. The DNA-APNH interaction is time, temperature, pH, concentration, and ionic strength dependent; optimal binding is attained at physiological ionic strength. The association reaction is rapid with $t_{1/2} < 30$ sec. In the presence of excess DNA added after the binding reaction reaches equilibrium, the DNA-APNH complex dissociates with 1st order kinetics ($k_b = 4 \times 10^{-4}$ sec^{-1}). Competition experiments with natural and synthetic DNAs indicate preferential binding of APNH to A-T rich and/ or single-stranded regions of the DNA. The single-strand preference is also corroborated by the effect of S1 nuclease treatment. The APNH fractionated by affinity chromatography on homologous and heterologous DNA-polyacrylamide columns shows lack of species-specificity in DNA binding. Based on these data and other preliminary results indicating destabilization of the DNA duplex by these proteins, it is hypothesized that the APNH may play an important role in transcription and/or replication.

Chromatin, the genetic apparatus of eukaryotes, is a supramolecular complex of DNA, RNA, histones and nonhistone proteins (NHCP). Ample evidence now accumulated shows that the regulated expression of genes in the differentiated state of eukaryotic cells results from the specific restriction of the genetic information in DNA by proteins of the chromatin. Positive correlation between quantitative, qualitative and metabolic changes of the NHCP and the modulations in the transcriptional capacity of the chromatin (see 1 for review of the evidence) set the stage for the first demonstration from the laboratory of J. Paul, later confirmed by many in a variety of systems, that it is the NHCP, and not the histones, that bestow

249

the specificity of template capacity on the chromatin (see 2 and 3 for reviews). Possible gene regulatory function for the NHCP implicit in these observations has provided the impetus for the current high interest in this class of proteins. It is clear, however, that the NHCP are extremely heterogeneous, many of them undoubtedly being enzymes of chromatin metabolism, and that proteins with gene regulatory functions may be only minor components of this total class.

For lack of an assay of a specific gene regulatory function many investigators have looked for NHCP that have specific affinity for DNA. The rationale for this approach is based on the operon paradigm of prokaryotes, which requires that gene regulatory proteins interact specifically with specific nucleotide sequences in the DNA. Various studies have shown that NHCP of different size and complexity bind to homologous and heterologous DNA (4,5) and low and high C_{ot} DNA (6), and that some of them exhibit species specificity in their ability to recognize DNA (7,8). In these studies the DNA-protein interaction was studied by methods of density gradient centrifugation or DNA-affinity column chromatography. Although these methods have been used successfully in demonstrating DNA binding of some specific regulator proteins (9,10), they are tedious and slow, and are particularly unsuitable for kinetic measurements of the binding reaction. Another procedure, which permits detection of DNA-protein complexes by their retention on nitrocellulose filters, has been most useful in detailed analysis of the lac repressor-operator interaction (11). Clearly, the molecular details of the NHCP binding to DNA also will have to be known in order to understand the molecular basis of the regulation of chromatin functions by these proteins.

In the studies reported here we have adapted the nitrocellulose filtration method to study the DNA binding properties of a selected class of NHCP (referred to as APNH) that was shown previously to bind DNA in vitro more efficiently than other nuclear acidic proteins (5). We have focused our research efforts on this particular class of proteins because it is considerably less heterogeneous than other groups of nuclear acidic proteins (12). This study describes the effect of various reaction conditions and the specificity of the interaction.

ISOLATION AND SOME PROPERTIES OF THE APNH

The term APNH has been used in our work to denote those NHCP that reconstitute and precipitate with DNA and histones

250

when a chromatin extract in buffered 2M NaCl-5M urea is dialyzed against 13 volumes of water to attain NaCl concentration
of 0.14M.   The isolation of APNH from the reconstituted deoxynucleoprotein and of other nuclear acidic proteins has been
described in detail elsewhere (5,12).   Briefly, purified rat
liver nuclei were washed extensively with 0.14M NaCl-0.05M
Tris HCl (pH 7.4) to extract the proteins of the nuclear sap
(NSP).   The nuclear material was then solubilized in 2M NaCl-
5M urea-0.01M Tris HCl (pH 8).   The residual material insoluble
in this high ionic strength solvent was removed by centrifugation at 40,000 x g for 60 min, and the viscous chromatin extract was dialyzed overnight against 13 volumes of distilled
water.   This resulted in the precipitation of a deoxynucleoprotein, leaving in the soluble phase greater than 90% of the
nonhistone proteins present in the chromatin extract which are
referred to as CAP-I.   The deoxynucleoprotein precipitate, containing DNA, histones and APNH, was solubilized in 2M NaCl-5M
urea-1mM phosphate buffer (pH 8) and fractionated by hydroxylapatite column chromatography in the presence of 2M NaCl-5M
urea.   The histones not retained by hydroxylapatite at 1mM
phosphate of the starting buffer were eluted in the breakthrough wash.   The APNH was then eluted at 0.05M phosphate; a
small amount of additional nonhistone proteins can be eluted
at 0.25M phosphate.   The NSP, CAP-I and APNH fractions contained variable amounts of nucleic acids which were removed
by ion exchange chromatography on DEAE-Sephadex in the presence of 8M urea (5).   To ensure that the results of nonhistone
protein binding to DNA are not influenced by minor contamination with histones or basic nonhistone proteins, the proteins
were passed also through Bio-Rex 70 columns in 0.4M guanidine-
HCl, 6M urea, 0.1M Na-phosphate (pH 7).

The APNH consists of less than 10% of the total NHP of the
chromatin extract and exhibits dramatically lower molecular
heterogeneity as compared to the NSP and CAP-I fractions (Figure 1).   The majority of proteins in this group appear as two
prominent bands with approximate molecular weights of 19,600
and 16,300 daltons, as determined by SDS-polyacrylamide gel
electrophoresis (12); a much smaller amount of higher molecular
size proteins appear to a slightly variable extent from preparation to preparation.   A composite of the CAP-I and APNH gel
patterns would be equivalent to the gel patterns of total NHCP
reported by others (e.g. 4,6,8).

The amino acid composition of the APNH has shown the proteins to be acidic (acidic/basic residue = 1.5) with no unusual
features that would explain their insolubility in common biochemical solvents (12).   Even in 2M NaCl about 50% of this

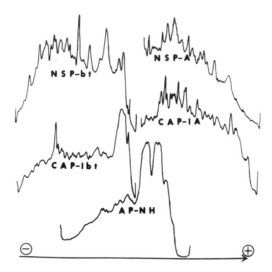

Fig. 1. Densitometric scans of SDS-acrylamide electrophoresis gels of the various nuclear protein fractions. See
text for explanation of abbreviations. The subfractions of NSP
and CAP-I were obtained as described in reference 5.

fraction is insoluble; complete solubility is attained only
with 5M urea, 2.5M guanidine-HCl or 0.1% SDS (unpublished data).
However, when mixed with DNA at NaCl and urea concentration of
2M and 5M, respectively, removal of NaCl and urea by dialysis
results in formation of DNA-APNH complexes that remain soluble
at low ionic strength. Using this procedure we have shown that
APNH binds more efficiently to DNA than NSP or CAP-I, and that
optimal interaction occurs at physiological ionic strength,
0.14M NaCl (5). It is interesting to note that the high affinity of APNH for DNA is also indirectly corroborated by data
reported by others. Prominent low molecular weight proteins
analogous to the major components of APNH can be seen in the
fraction of calf thymus NHCP that Allfrey et al. have reported
to elute in DNA-affinity chromatography at 2M NaCl-5M urea (6).
Gronow and Griffiths (13) have also shown analogous low molecular weight proteins that remain associated with DNA after
extraction of the bulk of nuclear nonhistone proteins, and that
can only be dissociated from DNA with 1% sodium dodecylsulfate.

APNH BINDING TO DNA ASSAYED BY NITROCELLULOSE FILTRATION

Technical Comments.

The theory and practical aspects of the nitrocellulose technique for the detection of DNA-protein complexes have been described by Riggs et al. (11) and Bourgeois (14). This technique is based on the fact that while pure DNA freely passes through the nitrocellulose filter, DNA-protein complexes are quantitatively retained on the filter. The extent of the interaction can be measured by employing isotopically labelled DNA. The number of DNA-protein complexes is proportional to the radioactivity retained on the filter. The fraction of the total input DNA retained on the filter is denoted R, where,

$$R = \frac{\text{cpm retained - cpm background free DNA}}{\text{Total cpm input}}$$

The sensitivity of this technique is a function of three variables: (1) the specific activity of the radioactive DNA, (2) the background radioactivity retained on the filter in the absence of protein, and (3) the relative DNA affinity of the protein under study.

Rat liver DNA was purified from citrate nuclei as described before (5) and banded in CsCl at 35,000 rpm for 48-60 hours. This DNA was labelled with ^{125}I by the method of Commerford (15) which yields DNA with a high specific activity. Background radioactivity retained by the nitrocellulose filter in the absence of protein is primarily a function of DNA purity. This is especially true of DNA labelled in the above method since ^{125}I-RNA, one of the products of this reaction, will be retained by the filter and give abnormally high backgrounds. For this reason, ^{125}I-DNA was banded again in CsCl to remove any contaminating RNA. Presoaking the nitrocellulose filters to 0.4M KOH for 20 min followed by rinsing and storing them in the binding buffer at 0-4° also helped lower backgrounds, as did the presence of 5% dimethylsulfoxide (DMSO) in the binding reaction. The background radioactivity retained by the filters in the absence of proteins was in the range of 0-5% of the input DNA. Salmon sperm DNA and E. coli DNA (Worthington Biochemicals) were also further purified by banding in CsCl. Poly d(A-T) and poly d(G-C) were procured from P. L. Biochemicals.

For a typical binding assay a small volume (∼10-20 μl) of APNH solution in 0.19M NaCl-5M urea-0.01M Tris HCl (pH 8) was rapidly mixed with 3-5 ml of a ^{125}I-DNA solution in the binding buffer and incubated at 25° for sufficient length of time to ensure that the reaction has reached equilibrium. The most commonly used binding buffer was 0.19M NaCl-0.01M Tris HCl (pH 7.4)-5% DMSO, but the composition of the binding reaction mixture and the conditions of the assay were modified according to the parameter under investigation. One ml aliquots were filtered on Schleicher and Schuell B6 membrane filters (25mm), followed by a wash with 1 ml of the binding buffer at 0-4°C. Care was taken during filtration to avoid complete drying of filters as this caused higher and erratic backgrounds. An aliquot of labelled DNA without protein was similarly filtered to determine the background. The filters were dried and counted in a scintillation spectrometer.

Binding Parameters.

With this binding assay we reconfirmed our previous observation, obtained by a different procedure (5), that the APNH fraction bound to DNA with greater efficiency than NSP and CAP-I fractions (Figure 2); bovine serum albumin showed no binding to

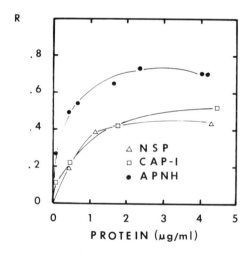

Fig. 2. Binding of various nuclear proteins to rat liver DNA. Various amounts of the proteins, in 0.19M NaCl-5M urea-0.01M Tris HCl (pH 8), were mixed with 3.0 ml of ^{125}I-DNA (0.08 µg/ml) in binding buffer, 0.19M NaCl-0.01M Tris HCl (pH 7.5)-5% DMSO, and the final volume was adjusted to 3.1 ml. Following 60 min incubation at 25°C, 1 ml aliquots were processed as described in the text to determine the fraction (R) of the input DNA bound to protein.

DNA under similar conditions (data not shown in Figure 2). The binding of NSP and CAP-I fractions has not been pursued, as yet, because of their high molecular heterogeneity. That they contain DNA binding proteins is not surprising, however, since proteins of all nuclear components must exist in a dynamic equilibrium. Furthermore, since the NHCP are synthesized in the cytoplasm (16), the different nuclear fractions will contain some of these proteins en route to the chromatin.

The binding of APNH to DNA depends upon time, pH, temperatue, ionic strength, monovalent cation, and protein concentration of the reaction mixture as shown below.

The association reaction is very rapid with maximal binding occurring within 2 min; $t_{1/2}$ of the reaction is less than 60 seconds (Figure 3). The reaction follows second order kin-

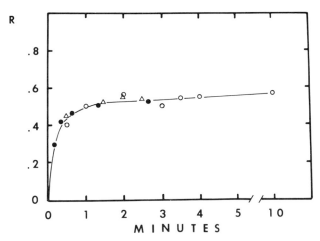

Fig. 3. Rate of association of APNH to rat DNA. Sufficient APNH to yield a final protein concentration of 2 µg/ml was mixed with 10 ml of ^{125}I–DNA (0.32µg/ml) in the binding buffer at 25°C. One ml aliquots were withdrawn at various intervals to determine R. Different symbols represent data from independent experiments.

etics. The fast rate of this reaction precludes kinetic measurements by the techniques of density gradient centrifugation and DNA-affinity chromatography that have been used previously for studying DNA-NHCP interactions. The nitrocellulose method allows these measurements because very dilute solutions of reactants can be used in the assay. The pH optimum of the

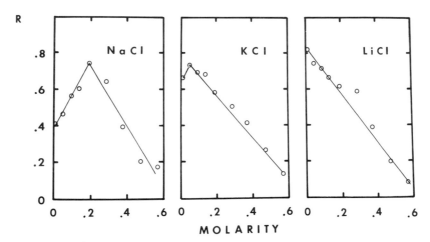

Fig. 4. Effect of the ionic strength and monovalent cations on the APNH binding to rat DNA. To 4 ml of 0.01M Tris HCl (pH 7.5) containing varying concentrations of either LiCl, NaCl or KCl, ^{125}I-DNA and DMSO were added to final concentrations of 0.08 µg/ml and 5%, respectively. One ml was filtered on a nitrocellulose disc to determine the background retention of DNA in the absence of protein for each salt concentration. Sufficient APNH to yield a final protein concentration of 1 µg/ml was added to the remaining 3 ml and the volume was adjusted to 3.1 ml. Following 60 min incubation at 25°C, 1 ml aliquots were processed to determine R.

reaction is in the range of 6.8 to 7.4. On either side of these values the extent of binding is reduced significantly.

The effect of monovalent cations on the reaction is shown in Figure 4. It can be seen that although the reaction clearly is dependent on the ionic strength, the actual value for maximal interaction varies with the type of cation employed. With Na^+ optimal binding is seen at 0.19M, while with K^+ and Li^+ this value is 0.05M and 0.01M, respectively. Decreased association with increasing ionic strength suggests that electrostatic interactions may be important in binding. We have also examined the effect of divalent cations Mg^{++}, Mn^{++}, and Ca^{++} and found them not to show any specific effects on the reaction.

The temperature effect on the binding reaction is only moderate in the 0-25° range, but the extent of reaction was markedly decreased between 25-80°C.

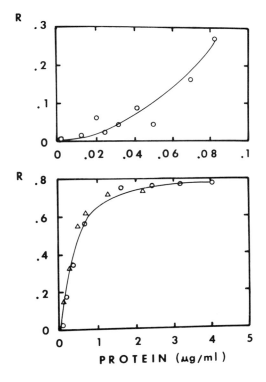

Fig. 5. Binding curve of APNH to rat liver DNA. Increasing amounts of APNH were mixed with 5 ml of ^{125}I-DNA (0.08 µg/ml) in the binding buffer, and the final volume was adjusted to 5.1 ml. Following 60 min incubation at 25°C, 1 ml aliquots were processed for determination of R.

The stoichiometry of the APNH-DNA binding is shown in Figure 5. The sigmoidal shape of the binding curve at low APNH concentrations suggests cooperativity of binding. Alternatively, sigmoidal shape might also be seen if more than one protein per DNA is required for retention of the DNA on the membrane, or if the binding proteins are required to be in a multimeric form with the monomeric form unable to bind DNA (14). The latter alternative requires that equilibrium establish rapidly between monomeric and multimeric forms of the binding protein. At this stage in our study of APNH we are unable to favor any explanation of the sigmoidal shape. The data show maximal interaction at mass ratio of protein/DNA = ~ 12.

Dissociation of the APNH-DNA Complex.

The binding of APNH to DNA under the conditions of the assay is reversible, as shown by the data of Figure 6. For this experiment ^{125}I-DNA (rat liver) and APNH were mixed and allowed to equilibrate; then at t=0, 200-fold excess of un-labelled rat liver DNA was added to the reaction mixture. At

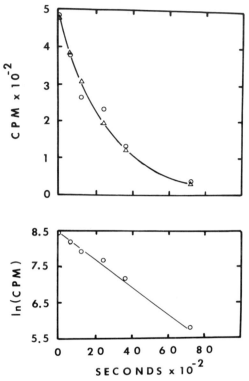

Fig. 6. Dissociation of the APNH-DNA complex. To 43 ml of ^{125}I-DNA (0.05 µg/ml) in the binding buffer 50 µl of APNH were added to yield a final protein concentration of 1.5 µg/ml. Following 90 min incubation at 25°C, five 1 ml aliquots were processed to determine the maximum binding. At t = 0, a 200-fold excess of unlabelled rat DNA was added to the reaction mixture. Thereafter, five 1 ml aliquots were withdrawn at 10 min intervals and processed for determination of the radio-active APNH-DNA complex remaining. The open circles represent experimental values; the open triangles represent calculated values assuming ideal first order kinetics and a first order rate constant, k_b, of 3.9×10^{-4} sec^{-1}.

various times afterwards, aliquots were withdrawn and filtered to determine the amount of ^{125}I-DNA APNH complex remaining. The complex dissociated with ideal first order kinetics. The rate constant of dissociation is calculated to be $3.9 \pm 0.97 \times 10^{-4}$ sec^{-1}. The half life of the complex under these conditions is 32 minutes. These results are in apparent contrast to our data from affinity chromatography of APNH on DNA-polyacrylamide column (17), which showed that a significant fraction of these proteins does not dissociate from DNA even with 2M NaCl-5M urea. However, in the affinity chromatography the proteins were in 5M urea, while in the nitrocellulose method the concentration of urea was diluted to less than 0.02M in the final assay mixture. The differences in the conformation of proteins under these conditions may account for the differences in their binding to and dissociation from DNA.

Specificity of the APNH-DNA Interaction.

The results discussed in this section were obtained from equilibrium competition type of experiments that were done as follows: Binding reaction mixtures were set up with a fixed amount of ^{125}I-rat liver DNA to which were added increasing amounts of unlabelled DNA to be tested for its capacity to compete for APNH binding. Equal amounts of APNH were added to each reaction, and after a time sufficient to allow maximum binding, the value of R was determined. Competition was reflected in the decreased retention of labelled rat liver DNA on filters. $D^{1/2}$ is defined as the concentration of the competitor DNA giving one-half of the maximum competition.

In our initial studies we employed this method to investigate the species specificity of APNH binding to DNA. To our surprise we found that rat liver APNH exhibited slightly greater affinity for salmon DNA than for homologous rat DNA (Figure 7A). When APNH was first passed through a salmon DNA-

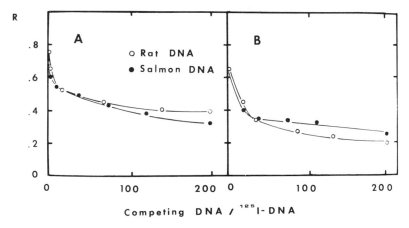

Fig. 7. Caption on next page.

polyacrylamide column at 0.14M NaCl-5Murea-0.01M Tris HCl (pH
8) to remove heterologous DNA binding proteins, the flow-
through APNH fraction showed only slightly greater affinity
for homologous rat DNA; however, this fraction still showed
significant binding affinity for salmon DNA (Figure 7B). Similar lack of absolute species specificity is also reported by
van den Broek et al. (4), who showed that rat liver NHCP passed
through E. coli and rat DNA-cellulose columns exhibited binding
to E. coli DNA. These results are different from those of
others (7,8) showing species specificity of some DNA-binding
nonhistone proteins and suggest that APNH may be distinct from
other NHCP with DNA-binding properties.

Equilibrium competition experiments comparing natural and
synthetic DNAs of varying G + C content and denatured and native homologous rat DNA suggest preferential binding of APNH to
A-T rich and/or single-stranded regions of the DNA. As shown
by the results in Figure 8 poly d(A-T) is the best competitor

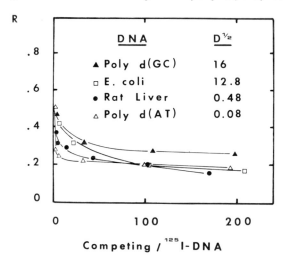

Fig. 8. Competition
of unlabelled natural
and synthetic DNA with
[125]I-rat DNA. Experi-
mental procedure was
similar to that des-
cribed for Figure 7,
except that total APNH
was used. $D^{1/2}$ is defined as the concen-
tration of the compet-
itor DNA giving one
half of the maximum
competition.

Fig. 7. Competition of unlabelled rat liver and salmon sperm
DNA with ^{125}I-rat liver DNA. Increasing amounts of unlabelled
rat or salmon DNA were mixed with ^{125}I-rat DNA (0.08 μg/ml) in
a final volume of 5.1 ml of the binding buffer, and 25 μl of
APNH were added to each tube to a final protein concentration
of 1 μg/ml. Following 60 min incubation at 25°C, 1 ml aliquots
were processed for determining R. Panel A: Total unfractionated APNH. Panel B: The APNH in 0.14M NaCl-0.01M Tris HCl (pH
8) was passed through a salmon DNA-polyacrylamide column to remove salmon DNA-binding proteins (17), and the breakthrough
proteins were used for the competition experiment.

while poly d(G-C) is the poorest competitor for the binding reaction. Although the only natural competitor DNAs shown in this experiment are from rat and E. coli, many other DNAs have been tested (results not shown here), and the results show that the extent of APNH binding correlated well to the A+T content of the DNA. The $D^{1/2}$s for the experiment in Figure 8 were 0.08, 0.48, 12.8 and 16 µg/ml competitor DNA for poly d(A-T), rat DNA, E. coli DNA, and poly d(G-C), respectively.

The greater affinity of APNH for single-stranded DNA is shown in Figure 9A. The $D^{1/2}$s for single- and double-stranded DNA were calculated to be 0.1 and 0.48 µg/ml competitor DNA,

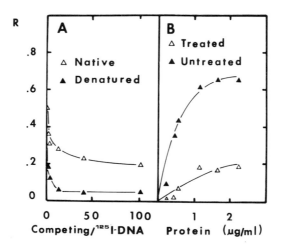

Fig. 9. Single-strand DNA preference of APNH binding. A: Native and heat denatured rat DNA was employed in competition experiments as described for Figure 7. B: Single-strand specific nuclease treated and untreated rat liver ^{125}I-DNA were used for binding according to the protocol described for Figure 5, except that the final volume was 3.1 ml. Treated DNA was prepared by incubating with the enzyme for 60 min at 37° in 0.05M Na-acetate (pH 5)-0.01M NaCl-0.001M ZnSO$_4$-5% glycerol and isolating the nuclease resistant ^{125}I-DNA duplex by banding in CsCl equilibrium gradient.

respectively. It is not clear, however, whether the observed competition by the native DNA reflects some protein binding to actual double-stranded regions, to some regions of single strandedness that may well exist in the native DNA preparations, or to A-T rich regions locally denatured by APNH. The single-strand preference is further supported by the effect of single-strand specific nuclease treatment of the ^{125}I-DNA prior to the binding assay under standard conditions. (Single-strand specific nuclease was a gift from R. Sinden). It can be seen that the maximal value of R attained at saturating APNH concentration is much lower for nuclease treated than the untreated DNA (Figure 9B). The low but significant APNH binding to the S_1 nuclease trimmed DNA might have resulted from incomplete nuclease action, from binding to actual double-stranded regions, or from APNH-catalyzed denaturation of A-T rich regions of DNA and subsequent binding to the resultant single strands. Preliminary experiments in our laboratory support the latter alternative, and current work is aimed at elucidating this question.

CONCLUSION

The proteins (APNH) described here are of a class distinct from other DNA-binding NHCP in their physico-chemical and DNA-binding characteristics. Therefore, it appears that some NHCP bind to DNA in a species-specific manner (7,8), while others bind preferentially to A-T rich and/or single-stranded regions of the DNA. Although it is now well established that the NHCP are important in dictating the specific template functions of the chromatin, the exact significance and biochemical consequences of their binding to DNA are not yet clear. One possibility suggested by our data on APNH binding to DNA is that these proteins may catalyze or facilitate local denaturation of the DNA duplex at A-T rich regions, an important step in the transcription and/or replication process. Preliminary experiments in our laboratory have shown that APNH enhance the transcriptional template activity of DNA in vitro in an E. coli RNA polymerase catalyzed reaction, and that they lower the melting temperature of synthetic poly d(AT) by about 35°C. Enhancement of DNA template activity (8,18,19) and helix-coil transition (20-23) catalyzed by DNA-binding proteins have been reported by others in various systems. In fact, the APNH are similar in many respects to the gene 32 protein (21) and DNA-unwinding proteins from calf thymus (Bruce Alberts, personal communication) that seem to play a role in maintaining a single-stranded configuration of the template in replication. We suggest that APNH may function in cooperation with the NHCP exhibiting species-specific binding to DNA for concerted regulation of the chromatin template functions.

Acknowledgement

This work was supported by U. S. Energy Research and Development Administration Research Contract No. AT (38-1) - 644.

References

1. V. G. Allfrey. In: "Acidic Proteins of the Nucleus", I. L. Cameron and J. R. Jeter, Jr. eds. Academic Press, New York (1974).

2. R. S. Gilmour. In: "Acidic Proteins of the Nucleus", I. L. Cameron and J. R. Jeter, Jr., eds. Academic Press, New York (1974).

3. G. Stein, T. C. Spelsberg and L. J. Kleinsmith. Science 183:817 (1974).

4. H. W. J. van den Broek, L. D. Nooden, S. Sevall and J. Bonner. Biochemistry 12:229 (1973).

5. G. L. Patel and T. L. Thomas. Proc. Nat. Acad. Sci. USA 70:2524 (1973).

6. V. G. Allfrey, A. Inoue, J. Karn, E. M. Johnson and G. Vidali. Cold Spring Harbor Symp. 38:785 (1973).

7. L. J. Kleinsmith, J. Heidema and A. Carroll. Nature 226:1025 (1970).

8. C. S. Teng, C. T. Teng and V. G. Allfrey. J. Biol. Chem. 246:3597 (1971).

9. M. Ptashne. Nature 214:232 (1967).

10. W. Gilbert and B. Müller-Hill. Proc. Nat. Acad. Sci. USA 58:2415 (1967).

11. A. D. Riggs, S. Bourgeois, R. F. Newby and M. Cohn. J. Mol. Biol. 34:365 (1968).

12. G. L. Patel. Life Sci. 11:1135 (1972).

13. M. Gronow and G. Griffiths. FEBS Lett. 15:340 (1971).

14. S. Bourgeois. In: "Gene Transcription in Reproductive Tissue", E. Diczfalusy, ed. Bogtrykkeriet Forum, Copenhagen (1972).

15. S. L. Commerford. Biochemistry 10:1993 (1971).

16. G. S. Stein and R. Baserga. Biochem. Biophys. Res. Commun. 41:715 (1970).

17. T. L. Thomas, M. A. Allen and G. L. Patel.   J. Cell Biol. 59:345a (1973).

18. M. Shea and L. J. Kleinsmith.   Biochem. Biophys. Res. Commun. 50:473 (1973).

19. N. C. Kostraba, R. A. Montagna and T. Y. Wang.   J. Biol. Chem. 250:1548 (1975).

20. A. P. Phillips.   Biochem. Biophys. Res. Commun. 30:393 (1968).

21. B. M. Alberts and L. Frey.   Nature 227:1313 (1970).

22. Y. Hotta and H. Stern.   Nature New Biol. 234:83 (1971).

23. B. M. Alberts and L. Frey.   J. Mol. Biol. 68:139 (1972).

Note added in proof

Recently, Sevall et al. (Biochemistry 14:782, 1975) have also described the binding of nonhistone chromatin proteins from rat liver to DNA, assayed by the nitrocellulose filtration method. Their results show that the binding specificity for the DNA is modulated by the ionic strength, and suggest preferential interaction between a subclass of nonhistone proteins and a subset of middle repetitive DNA sequences.

USE OF DNA COLUMNS TO SEPARATE AND CHARACTERIZE NUCLEAR NONHISTONE PROTEINS

Vincent G. Allfrey, Akira Inoue and Edward M. Johnson
The Rockefeller University
New York, New York 10021

Abstract

Methods have been developed for the covalent attachment of
single- and double-stranded DNA molecules to solid supports
such as Sepharose 4B, aminoethyl-Sepharose 4B, Sephadex G-25
and Sephadex G-200. Such columns have been used for the study
of the DNA-affinities of nonhistone proteins prepared from the
nuclei of calf thymus lymphocytes. The DNA-bound proteins are
displaced by stepwise increments in the salt concentration of
the eluting buffer – to yield sets of proteins which differ in
complexity and size distribution. Highly reproducible patterns
of elution of nuclear nonhistone proteins are obtained from
replicate DNA columns, and the proteins eluted at a given salt
concentration emerge at the same salt concentration on rechromatography. The method has been applied to studies of the
specificity of protein-DNA interactions. Comparisons of the
elution profiles and electrophoretic patterns of nuclear nonhistone proteins eluted from parallel columns containing low
$C_{0}t$, intermediate $C_{0}t$, and high $C_{0}t$ DNA sequences reveal that
some of the proteins interact preferentially with particular
DNA sequences. Some proteins bind to the "moderately repetitive" sequences and others to "unique" DNA sequences. DNA-
affinity chromatography under these conditions does not inacti-
vate the functions of a number of nuclear proteins, as judged
by the retention of enzyme activities, and by the capacity of
some proteins to specifically bind cyclic nucleotides such as
cyclic AMP or cyclic GMP. Several nucleotide-binding proteins
are released when the DNA columns are treated with deoxyribonuclease. A highly purified histone Hl kinase binds specifically to the DNA of the species of origin. DNA-binding resides
in the catalytic, rather than the regulatory, subunit of this
enzyme. Studies of the composition and metabolism of lymphocyte nuclear nonhistone proteins in cells stimulated by Concanavalin A show that many of the proteins enter the nucleus from
the cytoplasm shortly after exposure to the mitogen. This
intracellular movement appears to be highly selective, in the

sense that not all nuclear proteins increase proportionately during gene activation.

The experiments to be described are concerned with the application of new techniques of DNA-affinity chromatography to the study of the DNA-binding properties and functions of nuclear nonhistone proteins. Emphasis will be placed on methods for the covalent attachment of single-stranded and double-stranded DNA molecules to solid supports such as Sephadex G-25, Sephadex G-200, Sepharose 4B and aminoethyl-Sepharose 4B.

Three applications of DNA-column chromatography are presented. The first deals with the identification of nuclear nonhistone proteins which differ in their affinities for DNA sequences of different C_0t value. The second employs DNA-affinity chromatography to fractionate and partially purify nuclear proteins which specifically bind cyclic AMP or cyclic GMP. The third analyzes the interactions between a highly-purified, cAMP-dependent histone kinase and DNA.

Finally, we will describe some changes in the composition and rate of phosphorylation of nuclear nonhistone proteins in lymphocytes stimulated by the mitogen, Concanavalin A. Particular interest attaches to the demonstration that there is a massive, yet selective migration of proteins from the cytoplasm to the nucleus at early stages of gene activation.

METHODS OF DNA-AFFINITY CHROMATOGRAPHY

Deproteinization of DNA

Total calf thymus DNA was dissolved at a concentration of 2 mg/ml in 40 mM 2-[N-morpholino]-ethane sulfonic acid, pH 6.0 (MES buffer) and solid $NaClO_4$ was added to obtain a final concentration of 1 M. The solution was deproteinized by vigorous shaking for 10 minutes with an equal volume of 24:1 (v/v) mixture of chloroform and isopentanol. After centrifugation at 2,000 x g for 5 minutes, the aqueous phase was collected and shaken with $CHCl_3$-isopentanol as before. The deproteinization step was repeated four times. The final DNA solution was dialyzed against 40 mM MES buffer, pH 6.0, to remove $NaClO_4$.

Shearing and Sizing of DNA Fragments

A typical DNA preparation used for affinity chromatography had an average chain length (before sonication) of 0.5 - 1.5 x 10^4 base-pairs, and was 89% double-stranded, as judged by hyperchromicity analysis. The total calf thymus DNA was sheared

266

and separated into fractions enriched in "repetitive" (low C_0t), "moderately repetitive" (intermediate C_0t), and "unique" (high C_0t) sequences by hydroxyapatite chromatography of reannealed strands at different C_0t values (1-3). Deproteinized DNA was dissolved in 0.12 M Na phosphate buffer, pH 6.75, at a concentration of 2 mg/ml. Aliquots of the solution (45 ml) were placed in ice-jacketed 100 ml beakers and subjected to sonication using a Sonifier Model W185 (Heat Systems-Ultrasonic, Inc., Plainview, N.Y.) with a 1/2" disruptor horn placed 1 cm from the bottom of the beaker. Sonication was carried out for 1 minute at 75 watts output, for 1.5 minutes at 88 watts, and for two 1 minute periods at 88 watts, allowing 1 minute cooling intervals between bursts. The solution was centrifuged at 10,000 x g for 30 minutes.

The average lengths of the DNA fragments were determined by electron microscopy (4) and the observed lengths were expressed in terms of base-pairs (5) and plotted against frequency. Most of the fragments ranged in length from 200-400 base-pairs. Double-strandedness of the sheared DNA was estimated as 89%.

Preparation of DNA Subfractions Differing in C_0t Value

DNA was denatured by heating at 100°C for 5 minutes and then partially renatured at 60° to a C_0t value of 6. (The C_0t values are conveniently calculated from the product: $\frac{\text{Absorbance (of denatured DNA)}}{2}$ x hours (3)). The solution was passed through a hydroxyapatite column (which had been pre-treated with boiling 0.14 M Na phosphate buffer, pH 6.75, for 20 minutes and equilibrated with 0.12 M Na phosphate buffer at 60° to reduce binding of single-stranded DNA molecules). Elution of unbound DNA was carried out in 0.12 M Na phosphate buffer at 60°, and the single-stranded DNA in the eluate was processed for further reannealing as described below. The adsorbed low C_0t DNA was then eluted in 0.4 M Na phosphate buffer. Its double-stranded content was estimated at 73%. The proportion of eluted "repetitive" DNA sequences to unadsorbed DNA was 39/61.

The DNA not adsorbed to the hydroxyapatite column was dialyzed against water, concentrated, adjusted to a phosphate buffer concentration of 0.12 M and heated to 100°C for 5 minutes. It was then reannealed at 60° to a C_0t of 225. The solution was fractionated on hydroxyapatite, removing unadsorbed DNA with 0.12 M Na phosphate buffer, and eluting the intermediate C_0t DNA with 0.4 M Na phosphate buffer. This fraction, containing sequences of C_0t values 6 to 225, represented

13% of the total calf thymus DNA. Its double-strandedness was estimated at 60%.

The unadsorbed DNA was lyophilized and redissolved at a final Na phosphate buffer concentration of 0.7 M. After heating at 100°C for 5 minutes, the DNA solution was maintained at 60°C for 4 days. It was dialyzed against 0.12 M Na phosphate buffer and fractionated on hydroxyapatite. DNA recovery was 99%, of which 92% was in the adsorbed fraction. This fraction comprised sequences of C_0t values 225 - 3.9 x 10^4. It represented 44% of the total calf thymus DNA. The double-strandedness of this high C_0t fraction was estimated at 83%.

Removal of Single-Stranded Regions of Calf Thymus DNA Fractions by Nuclease S1 Treatment

Nuclease S1 (Aspergillus oryzae) was prepared from Sanzyme R (CalBiochem, Inc.) by chromatography on DEAE-cellulose, essentially as described by Sutton (6). The final enzyme preparation had a specific activity of 2 x 10^4 units/mg protein as tested on single-stranded, ^{3}H-thymidine labelled E. coli DNA (6). Calf thymus DNA fractions were dialyzed against 0.1 M NaCl - 3 x 10^{-5}M $ZnSO_4$ - 0.03 M Na acetate, pH 4.5, and treated with 800 units of S1 nuclease per mg of DNA for 30 minutes at 50°C. The solution was chilled; 1/10 volume of 1.5 M NaCl-0.15 M Na citrate, pH 7.0, was added, and the DNA was deproteinized as described above. The DNA was dialyzed against 0.12 M Na phosphate buffer at pH 6.8 and applied to pre-washed, equilibrated hydroxyapatite columns at 60°. The columns were washed with 4 volumes of 0.12 M Na phosphate buffer before removing the double-stranded DNA in 0.4 M Na phosphate buffer at 60°. The DNA was finally dialyzed against 40 mM MES buffer, pH 6.0. Its hyperchromicity indicated a double-stranded content in excess of 96%.

Covalent Coupling of DNA to Solid Supports

Solid matrices employed for DNA coupling included the polymerized dextrans, Sephadex G-25 and Sephadex G-200 (Pharmacia, Ltd.), the agar derivative Sepharose 4B (Pharmacia), and aminoethyl-Sepharose 4B prepared according to Cuatrecasas (7). Each of these materials was swollen in H_2O, poured into columns, and washed with 10 volumes of 40 mM MES buffer, pH 6.0.

The reagent used for coupling was the water-soluble carbodiimide, 1-cyclohexyl-3-[2-morpholinoethyl]-carbodiimide metho-p-toluene sulfonate (CMC) (Pierce Chemical Co., Rockford, Ill.) (8,9).

For the preparation of double-stranded DNA columns, heat denaturation must be avoided, and low temperatures were employed for the coupling reaction, as follows: 120 DNA absorbancy units$_{260nm}$ in 3.6 ml of 40 mM MES buffer, pH 6.0, were mixed with 1.5 g of gravity-packed solid support and excess water was removed by evaporation under a warm air stream, keeping the temperature of the paste below 30°. Eighty mg of CMC in 1 ml of H$_2$O were added, and the paste was maintained at 45°-50° for 4 hours. The coupling reaction was repeated twice, using 80 mg portions of CMC dissolved in 1 ml of H$_2$O and adding 0.2 ml of MES buffer. The reaction was allowed to proceed for 7 hours at 45°-50° each time. Finally, 5 ml of cold 0.1 M NaCl-1 mM EDTA-10 mM Na phosphate buffer, pH 7.0, were added and the matrix was allowed to swell for 12 hours at 4°. After decantation of excess fluid, the slurry was suspended in 20 volumes of 0.2 M NaHCO$_3$ for 10 hours at 4° in order to hydrolyze any adducts formed between DNA bases and the coupling reagent.

The polymer-linked, double-stranded DNA was poured into a column and washed with 50 volumes of 0.1 M NaCl-1 mM EDTA-10 mM Na phosphate buffer, pH 7.0, to remove free DNA. The material was stored in 5 ml of the washing buffer at 4°, using 2-3 drops of CHCl$_3$ as a preservative.

Yields of this reaction, using high and low molecular weight DNA preparations and a variety of solid supports, are presented in Table I. These yields are calculated from the total amount of DNA used for the reaction, and the amount of DNA released from measured aliquots of the DNA-matrix by treatment with pancreatic deoxyribonuclease I.

For coupling single-stranded DNA molecules to solid supports, the rate of the reaction can be accelerated by raising the temperature. Equivalent amounts of DNA (120 A$_{260nm}$ units), dissolved in 3.6 ml of 40 mM MES buffer, pH 6.0, were heated to 100° for 5 minutes and rapidly chilled in an ice bath, prior to adding 1.5 g of gravity-packed solid support. Excess liquid was removed by evaporation in an air-stream and 80 mg of CMC were added in 1 ml of H$_2$O. The paste was heated at 100° for 7 minutes and then rapidly chilled to 0°. An additional treatment with the coupling reagent was carried out, using 80 mg of CMC in 1 ml of H$_2$O and adding 0.2 ml of MES buffer. After mixing, the paste was heated to 100° for 7 minutes and then chilled. Subsequent steps in the preparation of polymer-linked, single-stranded DNAs are identical to those described for double-stranded DNA columns. Yields for the coupling of single-stranded DNAs to a variety of solid supports are summarized in Table I.

TABLE I

Covalent Binding of Deoxyribonucleic Acid to Solid Supports

Percent Binding[a] of

Matrix	Low-molecular-weight DNA[b]		High-molecular-weight DNA[c]	
	Double-stranded	Single-stranded	Double-stranded	
Sephadex G-25	61±1[d]	53±11	64±2[e]	
Sephadex G-200	27±7	33± 6	64±6	
Sepharose 4B	45±6	47± 7	84±2	
Aminoethyl-Sepharose 4B	58±5	56± 3	87±5	

Footnotes: a. Yields are expressed as per cent of original DNA coupled to 1.5 grams of gravity-packed solid support.

b. DNA sheared to 200-400 base pairs.

c. Non-sheared DNA $0.5-1.5 \times 10^4$ base pairs in length.

d. Low-molecular-weight DNA binding to Sephadex G-25 in the absence of the coupling reagent (CMC) was only 2% for double-stranded DNA under these conditions. No binding of single-stranded DNA was observed in the absence of CMC.

e. High-molecular-weight DNA binding to Sephadex G-25 in the absence of the coupling reagent was 18±5%.

270

TABLE II

DNA Content of Covalently-Bound Solid Supports

Concentration of

| Matrix | Low-molecular-weight DNA[a] | | | | High-molecular-weight DNA[b] | |
| | Double-stranded | | Single-stranded[c] | | Double-stranded | |
	mg/ml[d]	mg/g[e]	mg/ml[d]	mg/g[e]	mg/ml[d]	mg/g[e]
Sephadex G-25	1.19	6.1	1.03	5.3	1.24	6.4
Sephadex G-200	-	18.8	-	23.0	-	44.7
Sepharose 4B	-	22.5	-	23.5	-	42.0
Aminoethyl-Sepharose 4B	1.58	29.0	1.53	28.0	2.37	43.5

Footnotes: a. DNA sheared to 200-400 base pairs.

b. Non-sheared DNA 0.5-1.5 x 10^4 base pairs in length.

c. Single-stranded calf thymus DNA of C_0t greater than 200.

d. Concentration expressed as mg DNA per ml of gravity-packed solid support.

e. Concentration expressed as mg DNA per gram (dry weight) of solid support.

Properties of DNA-Affinity Columns

The coupling of double-stranded and single-stranded DNAs to Sephadexes, Sepharose 4B and aminoethyl-Sepharose 4B provides a range of solid supports of different physical properties and DNA content. The DNA concentration is expressed in terms of the volume or the dry mass of the different supports in Table II.

The stability of the linkage between DNA and Sephadex G-25 was examined after prolonged storage at 4° in 0.1 M NaCl-1 mM EDTA-10 mM Na Phosphate buffer, pH 7.35, using $CHCl_3$ as a preservative. Less than 1% of the DNA originally present was lost after 45 days. The corresponding figure for DNA bound to aminoethyl-Sepharose 4B was 2.6%. Tests for DNA leakage from Sephadex and aminoethyl-Sepharose columns during affinity chromatography showed that losses were minimal (usually less than 0.3% of the DNA originally present) over the wide range of salt concentrations used for elution of the DNA-binding proteins.

Isolation of Calf Thymus Nuclei and Preparation of Nonhistone Protein Fractions

Fresh calf thymus tissue was finely minced with scissors and homogenized in 5 volumes of 0.32 M sucrose-3 mM $MgCl_2$-0.5 mM $CaCl_2$. The nuclei were purified by differential centrifugation and washed extensively with 0.25 M sucrose-3 mM $MgCl_2$-0.5 mM $CaCl_2$. The yield from 500g of fresh tissue was 95g (wet weight) of isolated nuclei.

The nuclei were extracted three times with 10 volumes of 0.14 M NaCl containing 0.01 M $NaHSO_3$ as a protease inhibitor. Two different procedures were then used for preparing nonhistone proteins from the nuclear residue. The first is a non-denaturing procedure which selects the nuclear proteins soluble in 0.35 M NaCl. The nuclear residue was suspended in 10 volumes of 0.35 M NaCl-1 mM 2-mercaptoethanol-50 mM Tris (HCl), pH 8.0 and stirred for 30 minutes at 4°. After centrifugation for 2 hours at 200,000 xg, the extract was applied to a Bio-Rex 70 column which had been equilibrated with the extracting buffer. The run-off peak containing the nonhistone proteins was concentrated to 1 mg protein/ml by ultrafiltration through a Diaflo PM 30 membrane (Amicon Corp., Lexington, Mass.).

In the second procedure, the nuclear nonhistone proteins were extracted in high yield with the aid of denaturants. The nuclear residue remaining after extraction with 0.14 M NaCl-

0.01 M NaHSO$_3$ was suspended in 10 volumes of 0.25 N HCl to remove the histones. After centrifugation at 2,000 xg for 7 minutes, the nuclear residue was extracted twice more with 0.25 N
HCl. The de-histonized nuclei were then extracted in 6 M urea-
0.4 M guanidine hydrochloride-0.1% 2-mercaptoethanol-0.1 M Na
phosphate buffer, pH 7.45, by a modification of the method of
Levy et al. (10). DNA was removed from the extract by prolonged
centrifugation at 200,000 xg. The guanidine hydrochloride
concentration was lowered to 0.25 M, and the solution was concentrated by ultrafiltration, using a Diaflo PM 30 membrane.
At this stage the extract contained about 80% of the protein
present in the dehistonized nuclei.

Despite repeated extractions in 0.25 N HCl, a small amount
of histone remains in the nuclei and is removed along with the
nonhistone proteins in the urea-guanidine hydrochloride extract.
The contaminating histones were removed by ion-exchange chromatography on Bio-Rex 70. Elution was carried out in 6 M urea-
0.1 M Na phosphate buffer, pH 7.4, containing stepwise increments in guanidine hydrochloride concentration. The early
eluted fraction was free of histones as determined by the absence of characteristic histone bands in polyacrylamide gel
electrophoretic patterns (11).

Isotopic Labeling of Nuclear Nonhistone Proteins by Reductive Methylation

In order to facilitate monitoring of proteins during DNA-
affinity chromatography, aliquots of the different nuclear extracts (each comprising about 1/10 of the total) were labeled
by reaction with ^{3}H-formaldehyde and reduced with sodium boro-
hydride by a modification of the method of Rice and Means (12).

After extensive dialysis against 0.35 M NaCl-1 mM 2-mer-
captoethanol-50 mM Na phosphate buffer, pH 7.2, the radioactive
proteins from the 0.35 M NaCl extract had a specific activity
of 3.7 x 10^5 cpm/mg. The nonhistone protein fraction extracted
from dehistonized nuclei in 6 M urea-0.4 M guanidine hydrochloride was dialyzed after reductive methylation against 6 M urea-
0.25 M guanidine hydrochloride-0.1 M Na phosphate buffer, pH
7.4. Its specific activity was 3.3 x 10^5 cpm/mg. (It can be
calculated that these radioactivities correspond to the reductive methylation of only 5-17% of the protein molecules present). The ^{3}H-labeled protein fractions were then combined with
the corresponding unlabeled fractions before affinity chromatography on DNA columns.

DNA-Affinity Chromatography of Nuclear Proteins Soluble in 0.35 M NaCl

In studies of the DNA-binding properties of nuclear non-histone proteins solubilized in 0.35 M NaCl, the salt concentration of the extract was first reduced by dialysis against 50 mM NaCl in a low-ionic-strength buffer containing 10% (v/v) glycerol-10 mM NaHSO$_3$-0.1 mM EDTA-1 mM 2-mercaptoethanol-10 mM Tris (HCl), pH 7.5 (Buffer A). After centrifugation at 15,000 x g for 20 minutes, a 2 ml aliquot containing 1 mg of protein was applied to a 4.5 ml Sephadex G-25 column containing 4 mg of double-stranded, non-sheared calf thymus DNA. (The column was previously equilibrated with Buffer A containing 50 mM NaCl). After application of the sample, the column was washed with 50 mM NaCl in Buffer A. The proteins were subsequently eluted by stepwise increments in the NaCl concentration of the eluting buffer, as indicated in Figure 1. The column was finally washed with 6 M urea-0.4 M guanidine hydrochloride-0.1 M Na phosphate buffer, pH 7.5, to remove a small amount (less than 1%) of tightly-adsorbed protein. Recovery of the proteins from DNA-Sephadex G-25 columns under these conditions was 99-100% complete.

Control experiments included tests for protein binding to CMC-treated Sephadex G-25 columns which had not been reacted with DNA. Less than 0.8% of the total 0.25 M NaCl-soluble protein fraction was adsorbed to the DNA-free solid support.

DNA-Affinity Chromatography of Nuclear Nonhistone Proteins Extracted in 6 M Urea-0.4 M Guanidine Hydrochloride

The preparation of nuclear nonhistone proteins in 6 M urea-0.4 M guanidine hydrochloride (10) has the advantage of high yield but introduces the complication of protein denaturation. For this reason, proteins prepared in this way were combined with DNA using conditions of renaturation (13,14) which have been shown to be effective in chromatin reconstitution experiments leading to the selective synthesis of globin messenger-RNA (15,16).

The ^{3}H-labeled proteins were mixed in a dialysis bag with the DNA-solid support in 10 mM Tris (HCl), pH 7.5 - 10% glycerol-10 mM NaHSO$_3$-1 mM EDTA-1 mM 2-mercaptoethanol (Buffer B) containing 2 M NaCl and 5 M urea. The slurry was subjected to a series of dialysis steps against 5 M urea - buffer B solutions of decreasing NaCl concentration, in the order: 1 M; 0.6 M; 0.3 M; 0.15 M; 0.05 M. It was then dialyzed against 50 mM NaCl in buffer B without urea.

The slurry was poured into a column and washed with 50 mM NaCl in buffer B.

In studies of the differential binding of nonhistone proteins to DNA-aminoethyl-Sepharose 4B columns of different C_0t values, elution was carried out by stepwise increments in the salt concentration of the eluting buffer B, beginning with 0.05 M NaCl and progressing through 0.10 M, 0.15 M, 0.20 M, 0.25 M, 0.30 M, 0.40 M, 0.60 M, 1.25 M and 2.0 M NaCl. A minimum of 8 bed volumes of the eluting buffer was used at each step. The column was then washed with 2.0 M NaCl-5 M urea in buffer B, and finally washed with 6 M urea-0.4 M guanidine hydrochloride. All fractions were monitored for radioactivity and for protein content. Fractions eluted at a given salt concentration were combined, concentrated by ultrafiltration, and analyzed by electrophoresis in 10% polyacrylamide gels containing 0.1% SDS as described by Teng et al. (17).

Rechromatography of Protein Fractions Differing in DNA Affinity

Protein fractions eluted at a given salt concentration were rechromatographed on DNA-aminoethyl-Sepharose 4B to determine whether each subfraction would elute from the second column at the same salt concentration. After "annealing" to the DNA-column, the ^{3}H-labeled nuclear proteins were eluted by stepwise increments in the salt concentration of the eluting buffer B, beginning with 0.05 M NaCl and progressing through 0.10 M, 0.30 M, 0.40 M, 0.60 M and 1.25 M NaCl. The fractions collected at each salt concentration were separately concentrated by ultrafiltration, and a portion of each was analyzed electrophoretically. The remainder of each fraction was "annealed" to a separate column of aminoethyl-Sepharose 4B by the usual procedure. The elution of proteins from each column at different salt concentrations was monitored by radioassay of the emerging protein peaks as shown in Figure 5.

TESTS AND APPLICATIONS OF DNA-AFFINITY COLUMN CHROMATOGRAPHY TO THYMUS NUCLEAR NONHISTONE PROTEINS

DNA-Binding by Nuclear Proteins Soluble in 0.35 M NaCl

Calf thymus lymphocyte nuclei were used as a source of nonhistone proteins for studies of their interactions with long, double-stranded calf thymus DNA molecules. After washing the nuclei with 0.14 M NaCl, a protein fraction was extracted in 0.35 M NaCl-1 mM 2-mercaptoethanol-50 mM Tris (HCl), pH 8.0. Basic proteins were removed from the extract by chromatography

275

on Bio-Rex 70, and the nonhistone proteins emerging in the run-off peak were concentrated and applied to a DNA-Sephadex G-25 column (using a DNA/protein ratio of 4/1). The column was washed with buffered 50 mM NaCl to remove non-adsorbed proteins (about 60% of the total). The DNA-binding proteins (about 37% of the total) were subsequently eluted in 0.35 M NaCl (Figure 1A), the concentration originally used for extraction of the proteins from the isolated lymphocyte nuclei.

The fact that the retarded proteins interact with DNA and not with the matrix of the solid support was shown in two ways. First, the proteins do not bind to Sephadex G-25 columns which had been treated with the carbodiimide reagent in the absence of DNA. Secondly, the proteins which adsorb to the DNA-Sephadex column at 50 mM NaCl are completely displaced when the column is treated with bovine pancreatic deoxyribonuclease I, even when the enzyme treatment is carried out at salt concentrations which would not suffice to elute the adsorbed proteins from control DNA-Sephadex columns (Figure 1B).

DNA-Affinities of Nuclear Proteins Extracted in 6 M Urea - 0.4 M Guanidine Hydrochloride

The nuclear nonhistone proteins were extracted from de-histonized nuclei in greater than 80% yield in 6 M urea - 0.4 M guanidine hydrochloride - 0.1 M Na phosphate buffer, pH 7.45 (10). After removing the DNA from the extract by prolonged high-speed centrifugation, and elimination of a small fraction of contaminating histones by chromatography on Bio-Rex 70, the ^{3}H-labeled nonhistone proteins were applied to DNA-Sephadex G-25 (as described in Methods).

A large proportion (about 54%) of the proteins did not adsorb to the DNA under these conditions and emerged in the run-off peak, but the rest were retained. About 38% of the nonhistone proteins were eluted in 0.35 M NaCl, and about 6% were displaced by 2 M NaCl (Figure 2A). Treatment of the DNA-Sephadex column with deoxyribonuclease I at low ionic-strength displaced all but 0.2% of the adsorbed nonhistone proteins (Figure 2B).

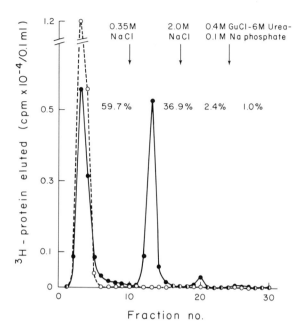

Fig. 1. A. Affinity chromatography of calf thymus lymphocyte nuclear nonhistone proteins on double-stranded calf thymus DNA covalently linked to Sephadex G-25. The proteins were extracted in buffered 0.35 M NaCl, labeled with ^{3}H by reductive methylation, and applied to the column in 50 mM NaCl. After washing the column with 50 mM NaCl to elute the nonadsorbed proteins, the salt concentration was raised to 0.35 M NaCl to elute most of the DNA-bound proteins. A small fraction of the latter was subsequently eluted in 2.0 M NaCl. The solid line shows the distribution of ^{3}H-labeled proteins emerging from the column of DNA-Sephadex G-25. The dashed line shows the corresponding elution diagram from Sephadex G-25 which had not been linked to DNA.

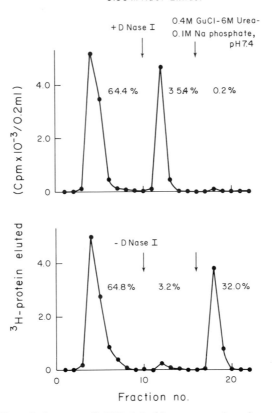

Fig. 1B. Release of DNA-binding proteins by deoxyribonuclease treatment of the DNA-Sephadex G-25 column. The 0.3 M
NaCl-soluble nuclear proteins, labeled with ^{3}H, were applied to
parallel columns of calf thymus DNA covalently linked to Sephadex G-25. After washing with 50 mM NaCl to remove the non-adsorbed proteins, one of the columns was treated with 1 mg of
bovine pancreatic deoxyribonuclease I in 2 ml of 10 mM NaCl in
Buffer A containing 2 mM Mg Cl_2 and 0.4 mM $CaCl_2$. Chromatography was continued using 50 mM NaCl. The upper panel shows
the release of the ^{3}H-labeled DNA binding proteins after DNAse
treatment, although 50 mM NaCl does not remove the DNA-bound
proteins from the control column (compare the elution profiles
of the upper and lower panels). The proteins attached to DNA
in the control column were subsequently released in 6 M urea-
0.4 M guanidine hydrochloride in 0.1 M Na phosphate buffer, pH
7.4.

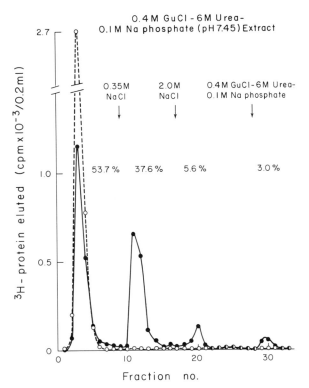

Fig. 2A. DNA-affinity chromatography of calf thymus nuclear proteins extracted from de-histonized nuclei in 6 M urea-0.4 M guanidine hydrochloride-0.1 M Na phosphate buffer. The proteins were labeled with ^{3}H by reductive methylation and combined with DNA under renaturing conditions as described in Methods. After washing the column in 50 mM NaCl to remove nonadsorbed proteins, the salt concentration was raised, as indicated, to elute the DNA-bound proteins. The solid line shows the distribution of ^{3}H-labeled proteins emerging from the column of double-stranded calf thymus DNA-Sephadex G-25. The dashed line shows the corresponding elution diagram from Sephadex G-25 which had not been linked to DNA.

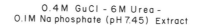

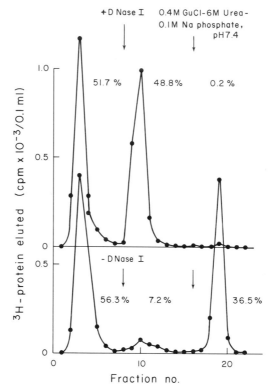

Fig. 2B. Release of DNA-binding proteins from DNA-Sephadex G-25 by deoxyribonuclease I treatment. The proteins extracted in 6 M urea-0.4 M guanidine hydrochloride-0.1 M Na phosphate buffer, pH 7.4, and labeled with ^{3}H, were applied to parallel columns of double-stranded calf DNA covalently linked to Sephadex G-25. After washing with 50 mM NaCl to remove non-adsorbed proteins, one of the columns was treated with DNAse I as described in the legend for Figure 1. The upper panel shows the release of the proteins after DNAse treatment. The lower panel shows the corresponding elution diagram from the control DNA column.

Concentration Dependence and Saturability of Protein-DNA Binding

The proportion of nuclear nonhistone proteins which adsorb to DNA-columns at low ionic strengths varies with the DNA/protein ratio. This is illustrated for proteins in the 0.35 M NaCl extract in Figure 3. The proteins were applied to double-stranded calf thymus DNA covalently linked to Sephadex G-25. The column was washed with 50 mM NaCl to remove unadsorbed proteins and the DNA-bound proteins were subsequently eluted with 6 M urea-0.4 M guanidine hydrochloride-0.1 M Na phosphate buffer, pH 7.45. The percent of the total protein bound to DNA is expressed as a function of the DNA/protein ratio in Figure 3.

It is clear that nonhistone protein binding to DNA is saturable; adding more DNA to the column does not lead to ever-increasing associations due to, e.g., electrostatic interactions between the proteins and DNA-phosphate groups.

Dependence of Protein-DNA Binding Upon Ionic Strength of the Medium

In the preceding experiments, the first interaction between nuclear proteins and DNA took place at low ionic strengths, usually in buffered 50 mM NaCl. Under these conditions, 35-40% of the nonhistone proteins were adsorbed to the DNA columns (Figures 1A, 2A). The proportion of the total nuclear protein which binds to DNA is lowered appreciably by raising the salt concentration. This is shown for proteins of the 0.35 M NaCl extract in Figure 4. In these experiments, the proteins were dialyzed against salt solutions ranging in NaCl concentration from 10 mM to 200 mM and applied to parallel DNA-Sephadex G-25 columns at a DNA/protein ratio of 4/1. After washing each column with 7 volumes of buffered NaCl of the appropriate molarity, the adsorbed proteins were eluted in 6 M urea - 0.4 M guanidine hydrochloride-0.1 M Na phosphate buffer, pH 7.45. The results (Figure 4) show that DNA binding is strongly dependent upon ionic strength. They also show that a substantial fraction (about 8%) of the proteins of the 0.35 M NaCl extract have sufficient DNA affinity to adsorb to the column at 0.2 M NaCl. Evidence that some of this binding is species- and sequence-specific is discussed below.

Reproducibility of Nuclear Protein Fractionation by DNA-Affinity Chromatography

The techniques outlined for the study of specificity in DNA-protein interactions fulfill two important requirements for

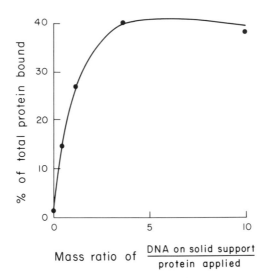

Binding of 0.35M NaCl-soluble nuclear
proteins to DNA-sephadex G-25 in 50mM
NaCl in buffer B, pH 7.5

Mass ratio of $\dfrac{\text{DNA on solid support}}{\text{protein applied}}$

Fig. 3. Effects of varying DNA/protein ratio on the binding of calf thymus nuclear proteins to calf thymus DNA–Sephadex
G–25. The proteins were extracted in buffered 0.35 M NaCl,
labeled with ^{3}H by reductive methylation, and applied to
double-stranded DNA–Sephadex G–25 columns in 50 mM NaCl. After
washing the column with 50 mM NaCl to remove unadsorbed proteins, the DNA-bound proteins were eluted and measured as described in Methods. The percentage of the protein bound increases until the DNA/protein ratio reaches 4/1.

Effect of NaCl concentration on binding
of 0.35M NaCl-soluble nuclear
proteins to DNA-sephadex G-25

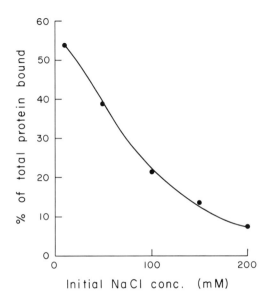

Initial NaCl conc. (mM)

Fig. 4. Effects of increasing ionic strength on the binding of nuclear nonhistone proteins to DNA-Sephadex G-25. The
nuclear proteins extracted in buffered 0.35 M NaCl were [3]H-
labeled by reductive methylation and applied to parallel columns of double-stranded calf DNA covalently linked to Sephadex
G-25, varying the NaCl concentration from 10 mM to 200 mM.
After washing to remove the unadsorbed proteins, the DNA-bound
proteins were eluted and measured. The per cent binding is
plotted as a function of the NaCl concentration during the
initial contact with the DNA column.

meaningful chromatography. First, the separation of nuclear
proteins according to DNA-affinity can be reproducibly achieved
by progressive elutions at increasing ionic strength. For
example, when the nuclear nonhistone proteins (extracted in 6
M urea-0.4 M guanidine hydrochloride-0.1 M Na phosphate buffer,
pH 7.45) are successively displaced from DNA-aminoethyl-Sepharose 4B columns using 0.10 M, 0.30 M, 0.40 M, 0.60 M and 1.25
M NaCl, the proteins in each fraction emerge at the same salt
concentration on rechromatography on a new DNA-aminoethyl-
Sepharose column. Figure 5 shows that - except for the forma-

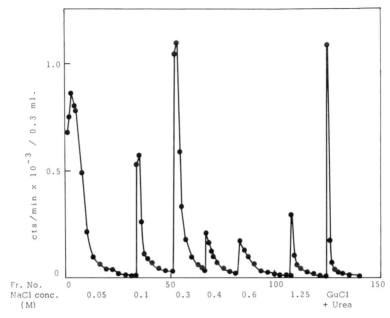

Fig. 5.   Rechromatography of nuclear nonhistone protein on
DNA–aminoethyl–Sepharose 4B.   The proteins were extracted from
de-histonized nuclei in 6 M urea-0.4 M guanidine hydrochloride-
0.1 M Na phosphate buffer, pH 7.45, and tritium-labeled by
reductive methylation.   They were applied to DNA-aminoethyl-
Sepharose 4B under renaturing conditions as described in Methods.   The proteins were then eluted by stepwise increments in
the NaCl concentration of the eluting buffer.
A (above).   Elution diagram of the nonhistone proteins emerging
from the first DNA-aminoethyl-Sepharose 4B column.   Each fraction was monitored for ^{3}H-activity which is plotted against the
salt concentration of the eluting buffer.
B (next page).   Rechromatography elution diagram for proteins
emerging from the first DNA-aminoethyl-Sepharose 4B column
in 0.1 M NaCl.   Note that, except for aggregate formation, the
proteins are again eluted in 0.1 M NaCl.
C (next page, bottom).   Rechromatography elution diagram for
proteins emerging from the first column in 0.4 M NaCl.

tion of some aggregates which are not displaced from the DNA
column unless denaturing agents are used - the elution characteristics of each sub-set of nuclear proteins are reasonably
constant.

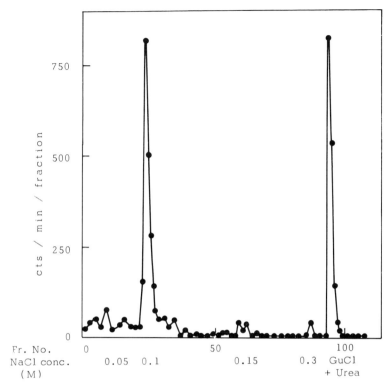

Fig. 5B (above)   Fig. 5C (below)

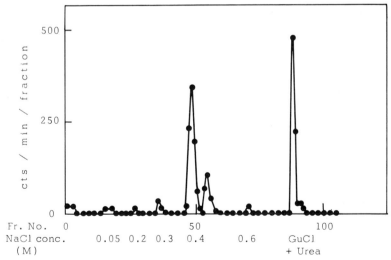

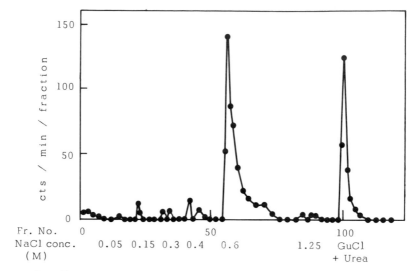

Fig. 5D. Rechromatography diagram for proteins emerging from the first DNA column in 0.6 M NaCl.

The second test for reproducibility compares the nature of the proteins eluted at a given salt concentration from dupli- cate DNA columns. Figure 6 compares the SDS-polyacrylamide gel patterns after electrophoretic analysis of the nuclear protein sub-sets eluted from different DNA-aminoethyl-Sepharose columns in 0.1 M, 0.4 M, and 0.6 M NaCl. High reproducibility of the complex protein banding patterns is obtained despite the fact that the experiments were carried out months apart on different preparations of nuclear nonhistone proteins (all extracted in 6 M urea-0.4 M guanidine hydrochloride-0.1 M Na phosphate buf- fer, pH 7.45) and on different preparations of DNA-aminoethyl- Sepharose 4B.

Evidence for Sequence-Specificity in Binding of Nonhistone Proteins to DNA

There are now numerous indications that nuclear nonhistone proteins interact selectively with DNA (e.g. 17-24). The rec- ognition of DNA sequences in higher organisms is complicated by the occurrence of large numbers of "repetitive" or similar DNA sequences (1,2,25). Such sequences can be separated from the remainder of the genome by shearing and heat-denaturation of the DNA followed by rapid reannealing of the multicopy strands and separation of the resulting duplexes by hydroxyapa- tite chromatography. Other DNA subfractions, representing less

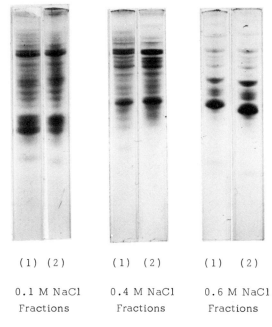

(1) (2)   (1) (2)   (1)   (2)

0.1 M NaCl   0.4 M NaCl   0.6 M NaCl
Fractions   Fractions   Fractions

Fig. 6. Tests for reproducibility of nonhistone protein
fractionation by affinity chromatography on DNA-aminoethyl-
Sepharose 4B. The proteins were extracted from de-histonized
nuclei in 6 M urea-0.4 M guanidine hydrochloride-0.1 M Na phosphate buffer, pH 7.45, and ^{3}H-labeled by reductive methylation.
They were applied to different columns of single-stranded, low
C_0t calf DNA. Elution was carried out by stepwise increments
in salt concentration as described in Methods. Sets of proteins emerging at each salt concentration were analyzed by
electrophoresis in 10% polyacrylamide-0.1% SDS gels at pH 7.4.
Comparison of the protein banding patterns obtained at each
salt concentration from different DNA-aminoethyl-Sepharose 4B
columns shows a very close correspondence. (Slight differences
in gel patterns at 0.4 M NaCl are due to differences in the
procedures employed for elution of this fraction in different
experiments).

common sequences, can be separated following more prolonged
reannealing and chromatography on hydroxyapatite columns (3).

Using these techniques to fractionate calf thymus DNA
(sheared to 200-400 base pairs), we have prepared subfractions
of widely differing C_0t values and then attached them to solid
support media. The binding affinities of calf thymus nuclear
nonhistone proteins for DNAs of different C_0t value, all derived from the same species, could then be compared.

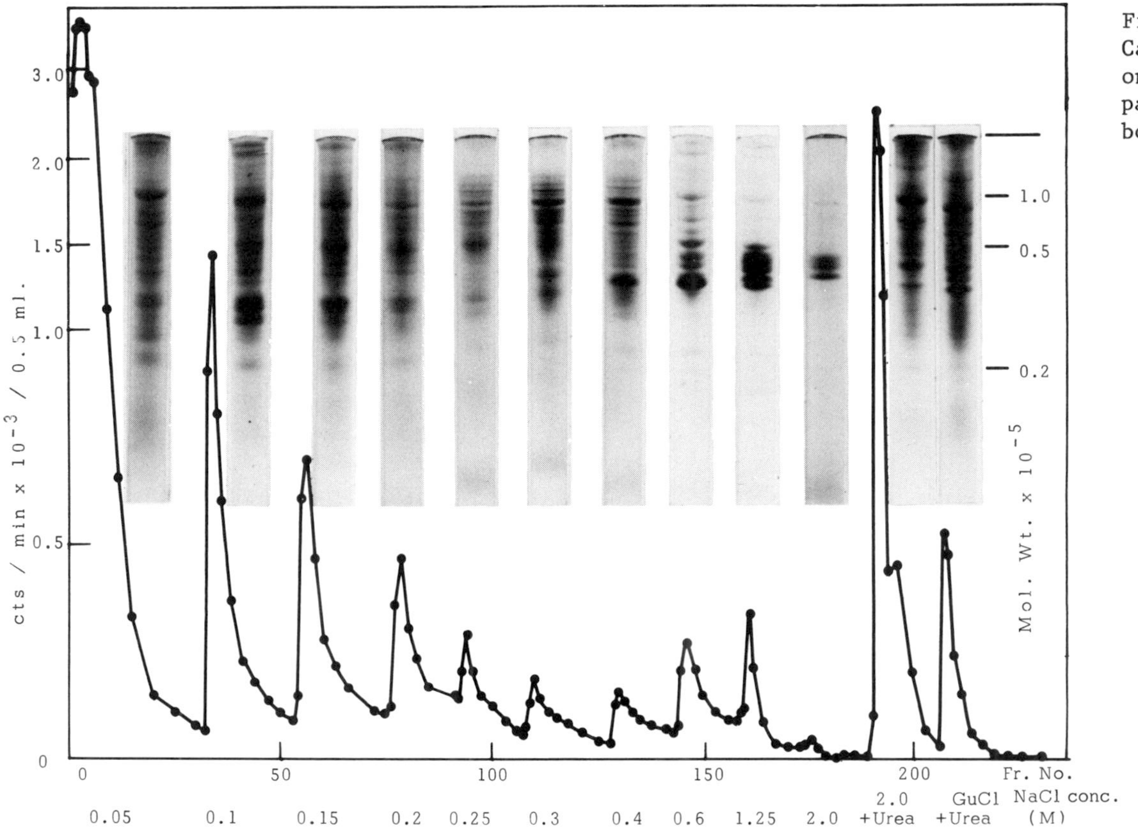

Fig. 7A.
Caption on next page at bottom.

288

The nonhistone proteins were extracted from de-histonized nuclei in 6 M urea-0.4 M guanidine hydrochloride-0.1 M Na phosphate buffer, pH 7.45, as described in Methods. Equal aliquots of the protein solution, labeled with ^{3}H by reductive methylation (12), were allowed to interact slowly with equal amounts of DNA-aminoethyl Sepharose 4B of different C_0t values, using the renaturing conditions employed for chromatin reconstitution experiments (13-16). Parallel columns were then poured, and the proteins were eluted by stepwise increments in the salt concentration of the eluting buffer, as shown in Figure 7. The gel electrophoretic pattern of proteins eluted at each salt concentration are indicated at the appropriate peak of the elution diagram.

The major conclusions can be summarized briefly: 1. Different nuclear nonhistone proteins adsorb to DNA with differing affinities. Some proteins bind so strongly that they may only be displaced at high ionic strengths, or by denaturing agents. 2. The protein sets eluted at different salt concentrations differ in their complexity and size distribution. 3. Comparisons of the electrophoretic patterns of proteins emerging at a given salt concentration from double-stranded high C_0t DNA (Figure 7A) or double-stranded low C_0t DNA (Figure 7B) show that they are not identical. Differences in the affinity of individual protein bands for high C_0t, intermediate C_0t, and low C_0t DNA sequences have been reproducibly observed. Several examples of differential protein binding as a function of C_0t value are presented in Figure 8.

Fig. 7A. Fractionation of nuclear nonhistone proteins from calf thymus on double-stranded, high C_0t (2.25 x 10^2 to 4 x 10^4) DNA covalently linked to aminoethyl-Sepharose 4B. The ^{3}H-labeled proteins were combined with DNA under renaturing conditions. The proteins were eluted with a discontinuous salt gradient and finally with 2.0 M NaCl-5 M urea and 6 M urea-0.4 M guanidine hydrochloride-0.1 M Na phosphate buffer, pH 7.45, each fraction being monitored as indicated for radioactivity. The proteins eluted at each salt concentration were analyzed by electrophoresis in 10% polyacrylamide-0.1% SDS gels. The patterns are shown for each protein set at the appropriate peak in the elution diagram. The corresponding molecular weight scale is given at the right of the figure.
B. Fractionation of nuclear nonhistone proteins on double-stranded, low C_0t (less than 6) DNA covalently linked to aminoethyl-Sepharose 4B.

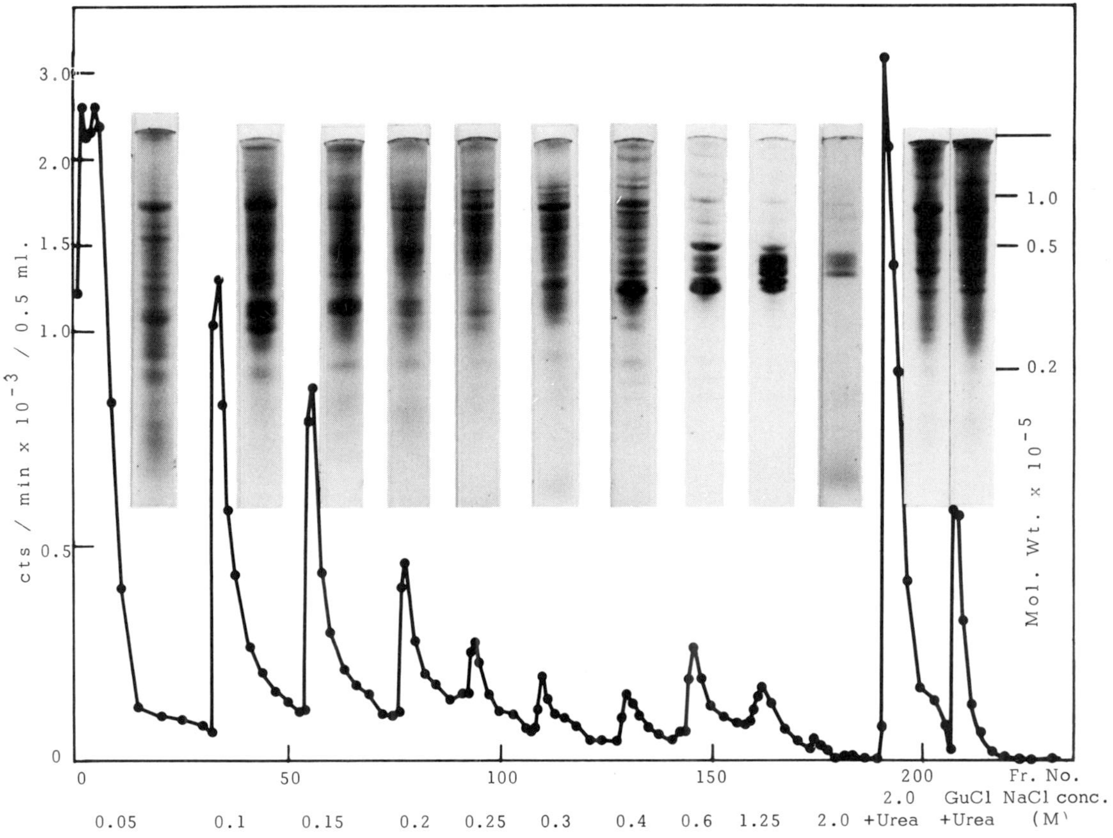

Fig. 7B
Caption on previous page at bottom.

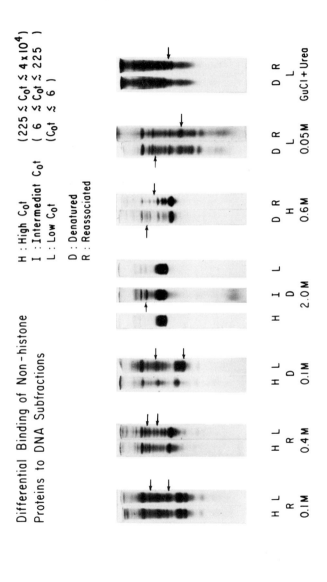

Differential Binding of Non-histone
Proteins to DNA Subfractions

H : High C_{ot} $(225 \lesssim C_{ot} \lesssim 4 \times 10^4)$
I : Intermediat C_{ot} $(6 \lesssim C_{ot} \lesssim 225)$
L : Low C_{ot} $(C_{ot} \lesssim 6)$

D : Denatured
R : Reassociated

H L	H L	H I L	H L	H L	D R	D R	D R
R	R	D	D	H	0.6M	L	L
0.1M	0.4M	0.1M	2.0M			0.05M	GuCl+Urea

Fig. 8. Comparisons of the electrophoretic banding patterns of protein sets eluted at corresponding salt concentrations from parallel columns of aminoethyl-Sepharose 4B with DNA subfractions differing in C_{ot} value or degree of strandedness. Side-by-side comparisons of proteins eluted at 0.1 M NaCl and at 0.4 M NaCl from double-stranded, high C_{ot} (H: 2.25×10^2 to 4×10^4) DNA and from double-stranded, low C_{ot} (L: less than 6) DNA are shown in the first two panels. Examples of differential binding of nuclear proteins to high C_{ot}, (continued on next page)

291

Analyses of the complete sets of electrophoretic patterns from parallel DNA columns differing in C_0t value show whether a given protein band interacts preferentially with "reiterated", "moderately repetitive" or "unique" DNA sequences. For example, protein bands of molecular weight 34,000 and 61,000 have greater affinity for double-stranded low C_0t DNA sequences and thus require higher salt concentrations for their displacement from the reiterated DNA sequences than from the unique sequences of the high C_0t DNA column. Conversely, a protein of molecular weight 50,000 binds more strongly to the high C_0t DNA sequences.

Similar comparisons of single-stranded DNAs differing in C_0t value also reveal a selective affinity of particular proteins for particular DNA sequences. Figure 8 includes an example of preferential binding of a nuclear protein of molecular weight 78,000 to intermediate C_0t DNA sequences.

The same techniques have been used to compare protein elution patterns from single-stranded and double-stranded DNA fractions of the same C_0t value. Clear indications of single- or double-strand specificity have been observed (Figure 8). The chromatographic behavior of some nuclear nonhistone proteins is consistent with the recognition of the physical state of the DNA as well as its nucleotide sequence.

Selective affinity based on DNA strandedness has been noted before. Proteins such as the lac repressor (26-28) and a restriction nuclease (29) are highly selective for double-stranded DNA. Conversely, the gene 32 protein of bacteriophage T4 (30) and the P8 protein of mouse fibroblasts (31) show affinity for single-stranded DNA but not for double-stranded DNA. Our experiments show that the nonhistone protein fraction from the lymphocyte nucleus includes representatives of both affinity classes.

Fig. 8 continued.
intermediate C_0t (I: 6-225), and low C_0t single-stranded DNA subfractions are shown in the third and fourth panels. The other panels illustrate differences in protein elution patterns from single-stranded and double-stranded DNA subfractions of the same C_0t value: R, reassociated DNA strands; D, denatured DNA strands. Major differences are indicated by the arrows. Smaller differences in relative concentrations are not indicated, but many exist.

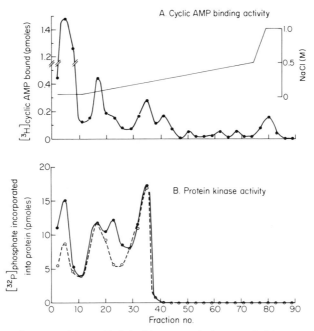

Fig. 9. Cyclic AMP-binding activity and histone Hl kinase activity of calf thymus nuclear nonhistone proteins separated by affinity chromatography on double-stranded calf DNA covalently linked to aminoethyl-Sepharose 4B. The proteins were extracted in 3 M NaCl and contaminating histones and DNA were removed (32). The proteins were applied to the DNA column in 0.05 M NaCl and eluted with a linear NaCl gradient from 50 mM to 500 mM NaCl, followed by stepwise elution in 1 M NaCl. Aliquots of 1 ml fractions were assayed for cyclic AMP-binding activity and for protein kinase activity (32). The upper panel shows the distribution of cyclic AMP-binding proteins in the eluate from the DNA column. The lower panel depicts the distribution of protein kinase activity as assayed in the presence (solid line) or absence (dashed line) of 5×10^{-6} M cyclic AMP. Note the different DNA-binding properties of different cyclic AMP-binding proteins. Also note that some peaks of cyclic AMP-binding activity do not coincide with peaks of protein kinase activity.

DNA-Affinity Chromatography of Cyclic Nucleotide-Binding Nuclear Nonhistone Proteins

Calf thymus nuclei contain specific cyclic AMP-binding proteins which are not removed from the nuclei in 0.14 M NaCl, but which can be eluted at high salt concentrations (e.g., 2 M NaCl)(32). After elimination of the histones by chromatography on Bio-Rex 70, the nuclear acidic proteins bind cyclic AMP with a specific affinity of 1.1 pmoles of cyclic AMP bound per mg of protein. In contrast, no cyclic AMP binding to histones has been observed (32). In order to determine whether cyclic AMP-binding proteins also interact with DNA, the nonhistone proteins were subjected to affinity chromatography on double-stranded calf thymus DNA covalently linked to aminoethyl-Sepharose 4B. The proteins were eluted with a linear salt gradient from 50 mM to 500 mM NaCl, and the column was finally washed with 1 M NaCl. Aliquots of each fraction were assayed for cyclic AMP-binding activity and for protein kinase activity (using histone H1 as substrate) as described in detail elsewhere (32).

The elution profiles of cyclic AMP binding activity and of histone H1 kinase activity are compared in Figure 9. It can be seen that there are a multiplicity of cAMP-binding peaks which are retained by the DNA column. Many of these peaks do not coincide with the observed positions of elution of either cAMP-dependent or cAMP-independent histone kinase activities (Figure 9, lower panel). DNAse treatment of the column releases the cAMP-binding proteins.

Similar studies of nuclear nonhistone proteins have indicated the presence of a number of cyclic GMP-binding proteins with strong affinities for double-stranded DNA columns (33). The possibility that such proteins which interact both with DNA and cyclic nucleotides in the lymphocyte nucleus may play a role in the control of transcription - in a manner analogous to the function of the cyclic AMP-receptor protein in the control of transcription of the lac and gal operons of E. coli (34-37) - warrants further investigation.

Specificity of DNA-Binding by a Histone H1 Kinase

A highly-purified histone H1 kinase - showing a 3 to 10-fold stimulation of activity in the presence of 5×10^{-6} M cyclic AMP - has been prepared from calf thymus tissue and chromatographed on double-stranded calf thymus DNA covalently attached to aminoethyl-Sepharose 4B. The elution of kinase activity is plotted as a function of the salt concentration of

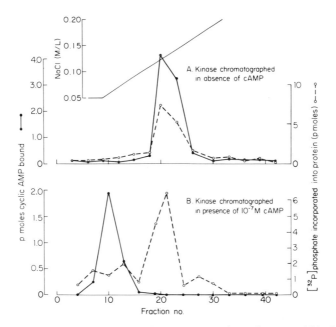

Fig. 10. DNA-affinity chromatography of a purified cyclic AMP-dependent protein kinase from calf thymus lymphocytes. The kinase was chromatographed on double-stranded calf thymus DNA covalently linked to aminoethyl-Sepharose 4B. Fractions emerging from the column during elution in a linear NaCl gradient were assayed for kinase activity (measured in the presence of 5×10^{-6} M cyclic AMP) (dashed line) and for cyclic AMP-binding activity (solid line). The upper panel shows that both activities emerge simultaneously from the DNA column if the enzyme is not dissociated into its regulatory and catalytic subunits by the addition of cAMP. The lower panel shows that the subunits differ in their DNA-affinities, the regulatory subunit emerging in the run-off peak while the catalytic subunit remains bound to the DNA.

the eluting buffer in the upper panel of Figure 10. The lower panel of Figure 10 shows the effects of dissociation of the holoenzyme into its regulatory (R) and catalytic (C) subunits. The C subunit binds to the DNA column while the R subunit does not.

Although the C subunit, as a basic protein, would be expected to interact with DNA, recent evidence indicates that the binding is not simply the result of electrostatic interactions. Using filter-binding assays to measure the formation of com-

295

plexes between ^{125}I-labeled calf thymus DNA and the purified protein kinase, it was found that DNAs from other species did not compete as effectively as did unlabeled calf DNA for kinase binding (32). The purified enzyme exhibits a clear preference for the DNA of the same species. The reason for this selectivity is not yet clear, but it should be noted that many nuclear nonhistone proteins, as well as histones, are subject to phosphorylation in the nucleus. The binding of kinases to specific DNA sequences in chromatin would provide an efficient mechanism for the in situ modulation of structure of other DNA-binding proteins.

SELECTIVE MIGRATION OF NONHISTONE PROTEINS FROM CYTOPLASM TO NUCLEUS AT TIMES OF GENE ACTIVATION

The preceding experiments have emphasized the interactions between lymphocyte nuclear proteins and DNA. What happens to these interactions when there is a massive activation of nuclear functions? In approaching this question, we have focused on changes in nuclear protein composition and metabolism in lymphocytes stimulated to divide by plant mitogens such as phytohemagglutinin and Concanavalin A (Con A).

Much of this work has been described elsewhere (38) but an important point deserves re-emphasis. The activation of the lymphocyte nucleus is accompanied by a massive migration of pre-existing cytoplasmic proteins into the nucleus. Cycloheximide inhibits the synthesis of nuclear acidic proteins by more than 90% but does not appreciably diminish the Con A-induced increase in the amount of protein entering the lymphocyte nucleus (38).

The relative proportions of the nuclear nonhistone proteins are altered at early stages of gene activation by Concanavalin A. Figure 11 compares the electrophoretic patterns of the phenol-soluble nuclear nonhistone proteins recovered from equal numbers of stimulated and non-stimulated lymphocytes. The striking increase in nuclear protein concentration is evident from the comparative intensities of the dye-binding patterns at the bottom of the figure. However, careful examination of the densitometric tracings shows that some proteins (e.g., proteins of molecular weights 43,000 and 150,000) increase in amount, while others (e.g., a protein of molecular weight 57,000) remain unaltered. This is a strong indication that highly selective mechanisms regulate the flow of nonhistone proteins from cytoplasm to nucleus. The importance of such controls in the feed-back control of nuclear activity cannot be over-emphasized.

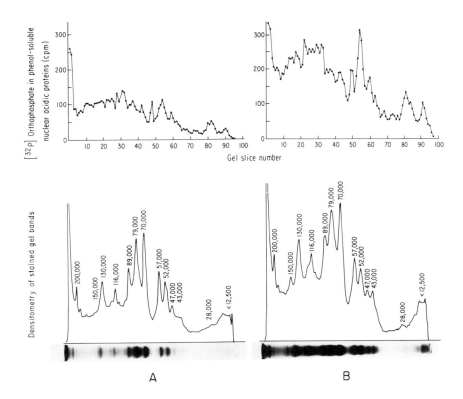

Fig. 11.   Effects of Concanavalin A on the composition and phosphorylation of lymphocyte nuclear nonhistone proteins.   After culturing for 2 hours in the presence or absence of Con A, lymphocytes were pulsed for 30 minutes with ^{32}P-orthophosphate. The nuclear proteins were isolated and analyzed by electrophoresis in SDS-polyacrylamide gels.   Each gel received nuclear protein from 2.5 x 10^6 lymphocyte nuclei.
A. ^{32}P-phosphate distribution (top) and densitometric tracing (bottom) of a single electrophoretic gel of nuclear proteins from control lymphocytes.
B. ^{32}P-phosphate distribution (top) and densitometric pattern (bottom) of nuclear proteins from lymphocytes incubated for 2 hours with Con A.   Panels have been aligned to permit vertical comparison.   Note that many nonhistone proteins increase in relative concentration in the nuclei of Con A-treated cells, but there is no significant change in the concentration of the protein banding at molecular weight 57,000.   There is also a preferential stimualtion of phosphorylation of a protein band of molecular weight 52,000 in the presence of Con A.

The upper panels in Figure 11 compare the distribution of ^{32}P-phosphate in the nuclear nonhistone proteins of control and Con A-stimulated lymphocytes. Within 2 hours of Con A addition, there is a 1.7-fold stimulation of ^{32}P-uptake into the nuclear proteins (on the average) but individual proteins differ greatly in their responses. Moreover, some of the proteins entering the nucleus in cells stimulated by Con A are not highly phosphorylated species. Increases in the relative amounts of proteins of molecular weights 43,000, 150,000 and below 12,500 are not accompanied by relative increases in ^{32}P-incorporation.

Thus, Con A activation of the lymphocyte leads to very rapid changes in the composition and metabolism of nuclear nonhistone proteins, as might be expected if some of these proteins play a direct role in chromatin structure and function. Which of these migratory proteins are DNA-binding proteins and how they are modulated by cyclic nucleotides and by protein kinases localized in the chromatin remain important topics for future investigation.

Acknowledgements

This research was supported in part by grants from the United States Public Health Service (GM 17383), the American Cancer Society (VC-114E), the National Foundation/March of Dimes (1-189), and the Rockefeller Foundation Program in Reproductive Biology. E.M.J. is a Special Fellow of the Leukemia Society of America.

References

1. R. J. Britten and D. E. Kohne. Science 161:529 (1968).

2. R. J. Britten and J. Smith. Carnegie Inst. Wash. Year Book 68:378 (1969).

3. D. E. Kohne and R. J. Britten. In: "Procedures in Nucleic Acid Research", Vol. II, G. L. Cantoni and D. R. Davies, eds. Harper and Row, New York (1971), p. 500.

4. W. Doerfler and A. K. Kleinschmidt. J. Mol. Biol. 50:579 (1970).

5. J. Eigner. In: "Methods in Enzymology", Vol. XII, Part B, L. Grossman and K. Moldave, eds. Academic Press, New York (1968), p. 386.

6. W. D. Sutton. Biochim. Biophys. Acta 240:522 (1971).

7. P. Cuatrecasas. J. Biol. Chem. 245:3059 (1970).

8. C. R. Astell and M. Smith. Biochemistry 11:4114 (1972).

9. D. Rickwood. Biochim. Biophys. Acta 269:47 (1972).

10. S. Levy, R. T. Simpson and H. A. Sober. Biochemistry 11: 1547 (1972).

11. S. Panyim and R. Chalkley. Arch. Biochem. Biophys. 130: 337 (1969).

12. R. H. Rice and G. E. Means. J. Biol. Chem. 246:831 (1971).

13. I. Bekhor, G. M. Kung and J. Bonner. J. Mol. Biol. 39:351 (1969).

14. R. S. Gilmour and J. Paul. FEBS Lett. 9:242 (1970).

15. T. Barrett, D. Maryanka, P. H. Hamlyn and H. J. Gould. Proc. Nat. Acad. Sci. USA 71:5057 (1974).

16. J. Paul, R. S. Gilmour, N. Affara, G. Birnie, P. Harrison, A. Hell, S. Humphries, J. Windass and B. Young. Cold Spring Harbor Symp. 38:885 (1974).

17. C. S. Teng, C. T. Teng and V. G. Allfrey. J. Biol. Chem. 246:3597 (1971).

18. L. J. Kleinsmith, J. Heidema and A. Carroll. Nature 226: 1025 (1970).

19. L. J. Kleinsmith. J. Biol. Chem. 248:5648 (1974).

20. H. W. J. van den Broek, L. D. Nooden, J. S. Sevall and J. Bonner. Biochemistry 12:229 (1973).

21. V. G. Allfrey, A. Inoue, J. Karn, E. M. Johnson and G. Vidali. Cold Spring Harbor Symp. 38:785 (1974).

22. J. S. Sevall, A. Cockburn, M. Savage and J. Bonner. Biochemistry 14:782 (1975).

23. J. D. Johnson, T. St. John and J. Bonner. Biochim. Biophys. Acta 378:424 (1975).

24. E. Jost, R. Lennox and H. Harris. J. Cell Sci. 18:41 (1975).

25. B. J. McCarthy and R. B. Church. Annu. Rev. Biochem. 39: 131 (1970).

26. W. Gilbert and B. Müller-Hill. Proc. Nat. Acad. Sci. USA. 58:2415 (1967).

27. A. D. Riggs, S. Bourgeois, R. F. Newby and M. Cohn. J. Mol. Biol. 34:365 (1968).

28. A. D. Riggs, H. Suzuki and S. Bourgeois. J. Mol. Biol. 48: 67 (1970).

29. R. Yuan and M. Meselson. Proc. Nat. Acad. Sci. USA. 65: 357 (1970).

30. B. M. Alberts and L. Frey. Nature 227:1313 (1970).

31. R. L. Tsai and H. Green. J. Mol. Biol. 73:307 (1973).

32. E. M. Johnson, J. W. Hadden, A. Inoue and V. G. Allfrey. Biochemistry, in press (1975).

33. V. G. Allfrey, A. Inoue, J. Karn, E. M. Johnson, R. A. Good and J. W. Hadden. In: "The Structure and Function of Chromatin", Vol. XXVIII (new series), CIBA Foundation Symposium (1975), p. 199.

34. I. Pastan and R. Perlman. Science 169:339 (1970).

35. G. Zubay, D. Schwartz and J. Beckwith. Proc. Nat. Acad. Sci. USA. 66:104 (1970).

36. M. Emmer, B. deCrombrugghe, I. Pastan and R. Perlman. Proc. Nat. Acad. Sci. USA. 66:480 (1970).

37. S. P. Nissley, W. B. Anderson, M. Gottesman, R. L. Perlman and I. Pastan. J. Biol. Chem. 246:4671 (1971).

38. E. M. Johnson, J. Karn and V. G. Allfrey. J. Biol. Chem. 249:4990 (1974).

Subject Index

A

Actin, 232
Actinomycin D
  effect on
    histone methylation, 144, 147
    nonhistone protein phosphorylation, 83
  in reverse transcriptase reactions, 5
S-adenosylmethionine as donor in histone
  methylation, 128
Alkaline sucrose gradient sedimentation, 5
Antibodies, to nonhistone proteins, 190
ATP, as phosphate donor, 47–48

B

BioRex 70 chromatography, 36
Blastula, nonhistone protein phosphorylation
  in, 83–86

C

Carcinogens, effect on histone methylation,
  136, 137, 138
Cell Cycle, 1–17, 59–65, 82–83
  biochemistry of, 1–2
  chromatin template activity, 2, 60–63
  deposition of newly synthesized histones
    during, 93–108
  gene regulation during, 1–2, 59–66
  histone phosphorylation, 119–124
  nonhistone chromosomal protein metabolism
    during, 2, 82
  nonhistone protein phosphorylation during,
    14, 49, 59–66, 82–83
  transcription of histone genes during,

10–15, 63–65
Cellular differentiation, 227
  nonhistone protein metabolism during,
    67–92
  nonhistone protein phosphorylation, 80–81
Cesium chloride gradient fractionation,
  25–26, 28–29, 30, 32, 94–108
Chromatin
  arrangement of
    globin genes in, 222–226
    proteins in, 213–226
  composition of, 214
  digestion with staphylococcal nuclease,
    214–226
  electron microscopy of, 214
  fine structure of, 222–226
  formaldehyde fixation of, 237–239
  fractionation of, 189–207, 227–246
  hormone receptors, 153–186
  isolation of, 189
  lysine methyltransferases in, 129
  nuclease digestion of, 234
  preparation of, 21, 94–95, 103, 156
  subunit structure of, 232–246
  template
  active, 229–246
  inactive, 228–246
Chromatin reconstitution, 2, 12–15, 20, 22,
  26–33, 35–41, 46, 60–66, 220–222,
  274–275
Chromatin structure, 213–226, 227–246
  genome packaging, 114, 213–226, 227–246
  effect of
    histone phosphorylation, 113–124
    nonhistone protein phosphorylation,
      45–53
  repeating units, 114, 219, 234
  role of
    histones, 114–124
    nonhistone proteins, 114, 154

A
B 5
C 6
D 7
E 8
F 9
G 0
H 1
I 2
J 3